GW00721818

bio-
TECHNOLOGY

bio- TECHNOLOGY

CHANGING THE WAY NATURE WORKS

JOHN HODGSON

CASSELL

Editor Mike March
Designers Niki Overy,
Frankie MacMillan
Picture Editor Alison Renney
Picture Researchers Rose Taylor,
Mary Fane

Design Consultant John Ridgeway
Project Director Lawrence Clarke

Advisors Bernard Dixon,
Steve Connor

Contributor John Hodgson

Computer generated molecular graphics by Chemical Design Ltd, Oxford, using Chem-X software.

AN EQUINOX BOOK

Planned and produced by:
Equinox (Oxford) Ltd
Musterlin House,
Jordan Hill Road
Oxford OX2 8DP

First published in the UK 1989 by
Cassell, Artillery House, Artillery Row,
London, SW1P 1RT

Distributed in Australia by
Capricorn Link (Australia) Pty Ltd
PO Box 665, Lane Cove, NSW 2066

British Library Cataloguing in Publication Data Hodgson, John
Biotechnology 1. Biotechnology I. Title
302.2

ISBN 0 304 31783 7

Cover pictures

Main picture: Cetus Corporation

Corner picture: Chemical Designs Ltd, Oxford

Printed in Spain by H. Fournier, S.A.

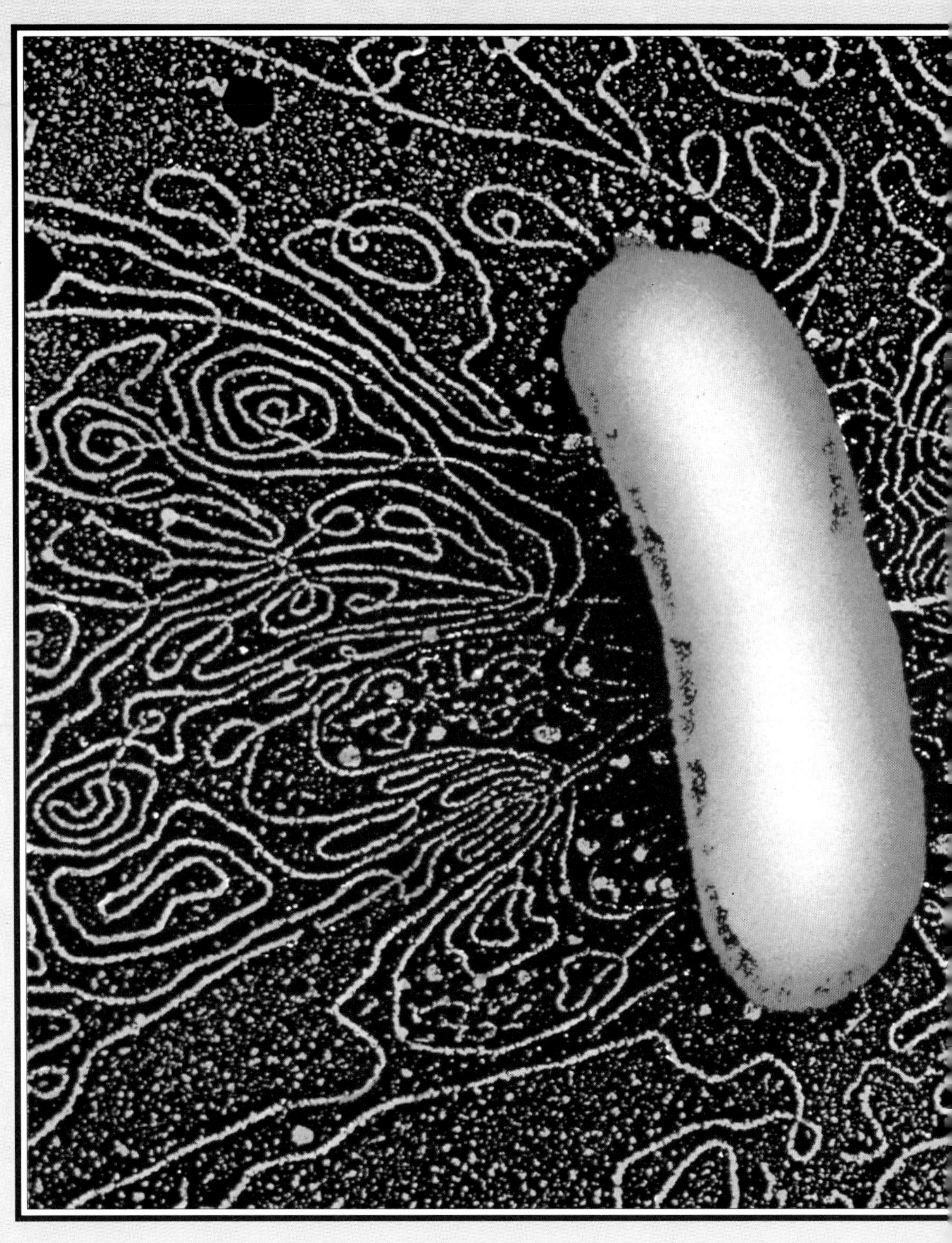

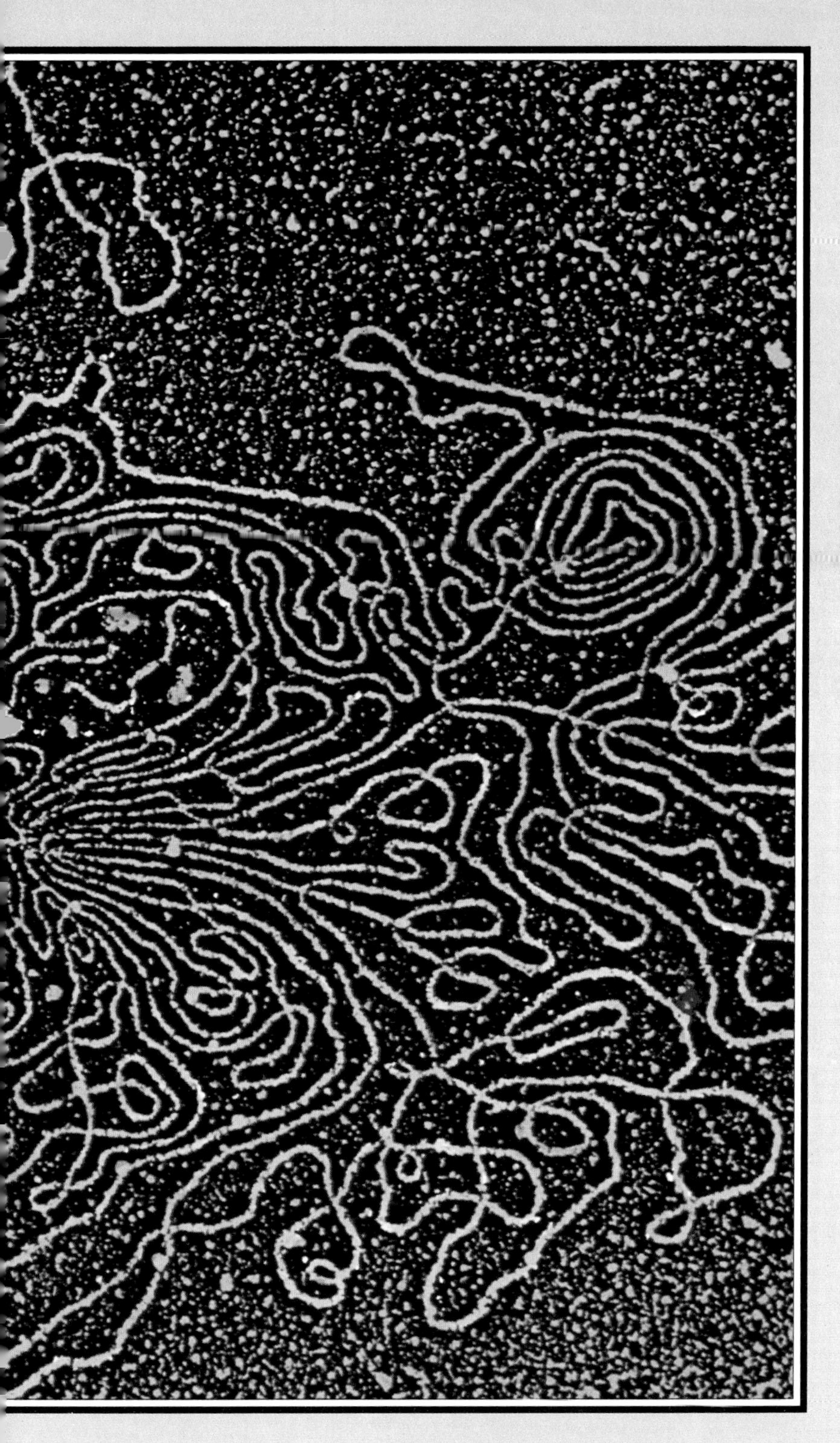

Contents

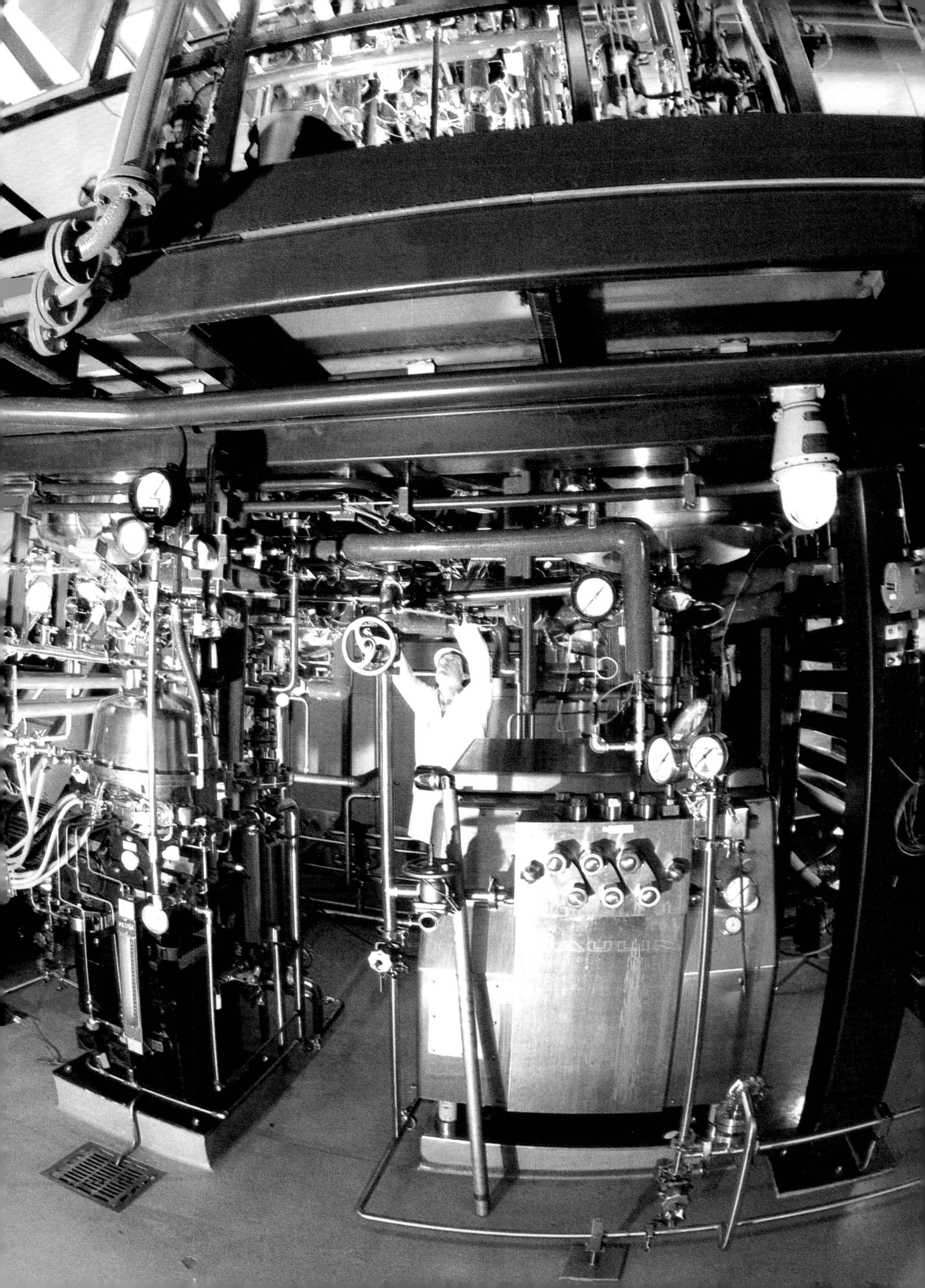

The Chemistry of Life

1

Cells and organisms...Cellular symbiosis... Differences between organisms...The genetic blueprint...Factors and dominance...Chromosomes and genes...DNA: composition and structure... Breaking the DNA code...The decoding machine... The mechanism of replication...Gene cloning... Restriction enzymes...Vectors

For many centuries scientists had wanted to find out what living matter was made of. Why were living organisms different from inanimate materials? What distinguished a dog from a trout or a bee from a human? Were there similarities between life forms which would show why they were alive?

All living matter is made up of tiny compartments, or cells, and only cells can give rise to cells. Cells are the fundamental units of life, its building blocks and its builders. Higher organisms such as plants, animals and human beings are composed of many millions of cells grouped in organs and performing different functions. Bacteria are self-contained single-celled microscopic organisms.

Cells vary in size from one or two millionths of a meter (1–2μm) in diameter in bacteria to around 10μm for most animal and plant cells. Some specialized cells are much larger. The egg cells of certain birds may exceed 10cm in diameter, while the nerve cells connecting a giraffe's brain to its toes are 4–5m long. Cells are complex microscopic bags of chemicals surrounded by a very thin flexible "skin", the cell membrane. This holds together the contents of the cell and controls the passage of nutrients into, and waste products out of, the cell. Most plant and microbial cells are strengthened by a rigid outer wall.

The most important property of all living cells is the ability to produce replicas of themselves from animate materials. Indeed, the capacity for self-perpetuation can be taken as the definition of life. Living matter can produce copies of itself without any instructions from outside. Lifeless matter cannot.

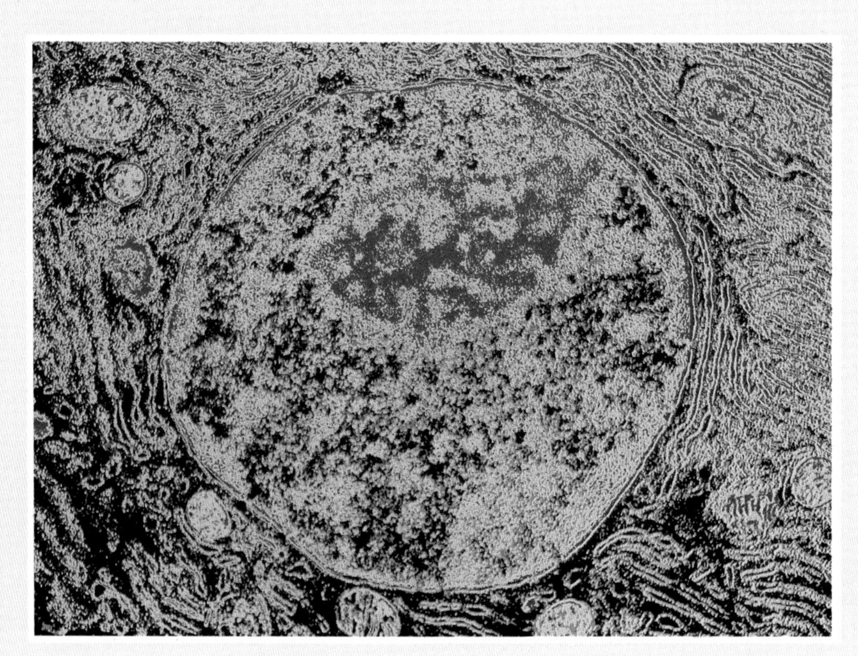

◀ One of life's building blocks – a cell from the pancreas of a bat, magnified about 10,000 times. The nucleus (central area) directs cell building, maintenance and activity. The cell is divided into compartments by a highly organized network of membranes.

Microorganisms and germs

Microorganisms are organisms that are too small to be seen with the naked eye. Germs are infectious microorganisms which cause disease. There are four main categories of microorganism – bacteria, fungi, viruses and parasites.

Bacteria are responsible for cholera, typhoid, bubonic plague, anthrax, whooping cough, tetanus, diphtheria, boils, gonorrhea and syphilis. More than 10,000 average bacteria could be placed on a dot measuring 0.5mm across. Under a light microscope, bacteria appear as tiny transparent spheres or lozenges which sometimes contain spores. Bacteria grow to a certain size and then multiply by dividing into two. Most bacteria do not cause disease and many are beneficial to humans. On perfectly healthy skin, there may be over 10 million bacteria per square centimeter. These prevent infections by harmful (pathogenic) organisms. Bacteria also convert milk to cheese and yoghurt, and produce many antibiotics, food flavorings – like soy sauce, monosodium glutamate and vinegar, and valuable chemicals.

Fungi (singular, fungus) are much larger and more complex than bacteria. Although individually they may be only a few thousandths of a millimeter in diameter, they often grow as long filaments. The molds seen on bread and the blue of blue cheese are filamentous fungi. Sometimes the filaments come together to form mushrooms or toadstools. Only 100 or so human diseases are caused by fungi, including relatively mild conditions such as ringworm and athlete's foot. A fungus called Penicillium *produces penicillin.*

Yeasts are special types of fungi which appear as globular balls or "ovoids". Yeasts cause some diseases (thrush, for instance) but are better known for their role in fermenting beer and wine.

Viruses are much smaller and simpler than either fungi or bacteria. It would take 100,000 large viruses to cover a punctuation point. Viruses invade other organisms and take over their growth processes to produce more viruses. They cause influenza, colds, hepatitis, yellow fever, AIDS (acquired immune deficiency syndrome), warts, smallpox, polio, rabies and many other human diseases. Many important plant and animal diseases (tobacco mosaic and foot-and-mouth disease, for instance) are viral in origin. There are also viruses that invade bacteria. They are called bacteriophages.

Parasites are a group of relatively complex organisms that cause disease in humans and animals, including malaria, sleeping sickness and scabies. The malaria parasite, Plasmodium falciparum, *has two hosts – people and mosquitoes – and has at least five distinct forms.*

Cells and organisms

Despite the almost inexhaustible variety of life, the cells of all organisms are remarkably similar in composition and function. Chemically, they are virtually identical. Oxygen, carbon, hydrogen, and nitrogen account for 95 percent of the cell components. Calcium, sodium, potassium, sulfur, chlorine, magnesium and other elements are also present in small amounts.

Cells from different organisms are also very similar biochemically. Scientists now know that there is very little difference between the central biochemical pathways (the series of reactions producing most of the cell's important building blocks) in organisms as diverse as man, the mouse, *Escherichia coli* (the gut bacterium) and spinach. Although only a few organisms have been studied in any great depth, it is believed that these similarities extend to all species.

The theory of "natural selection" propounded by Charles Darwin (1809–1882) was associated with the idea that the diversity of complex creatures on the Earth could have evolved from a single primeval organism. Organisms are similar because they have a common ancestry. For the first two billion years of the history of life, all organisms were *prokaryotes* (from the Greek for "before nucleus"). Like today's bacteria, these lacked obvious internal organization. Then, about 1.5 billion years ago, a quite momentous development occurred. The *eukaryotes* (from the Greek for "true nucleus") appeared. Their genesis is uncertain but they are distinct from the prokaryotes in that eukaryotic cells contain structural elements – organelles – that divide each cell into compartments.

Prokaryotes are usually single-celled. Most multicellular organisms including humans, animals and plants, insects, fungi and yeasts are eukaryotes. In these higher organisms, groups of cells cooperate. The tissues and organs of eukaryotes are organizations of cells. The cells of one organ perform different functions from those of other organs. In animals, for instance, the cells of the pancreas produce substances to help digest food. Liver cells store fats and produce bile. Brain cells transmit and receive chemical signals from the other cells in the body. Cells are responsible for the formation of bone and skin and hair and muscle. Other cells make antibodies, consume invading bacteria, or produce adrenalin.

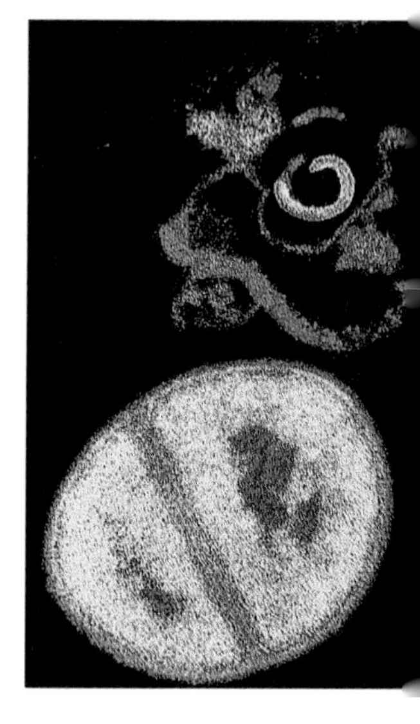

▶ Staphylococcus aureus, a bacterium, in the process of division. A double thickness of the rigid wall which surrounds the organism is developing internally, dividing the cell in two. After further growth of the wall, two spherical daughter cells separate. Each contains an identical set of genetic information and can exist independently. They may themselves divide half an hour after separating. This rapid growth helps Staphylococcus to cause boils and abscesses in humans. Antibiotics like penicillin destroy the cell wall, causing the organism to burst. The debris of a dead organism is seen at the top of the photograph.

The genetic blueprint

To build complex structures, the cell, like any engineer, needs both a plan and reliable construction equipment. The equipment is the cell and its components. The plan is the genetic information, which is the same whether it is in the cells of your brain, bile duct or big toe. An enormous number of different cell types can be fabricated by combining different elements of an all-purpose plan, even though similar biochemical construction machinery is used for each cell.

The genetic information determines the distinguishing features of all organisms. It controls the growth of cells and their development into tissues. Genetics, the study of the information content of cells, has had a profound impact on how scientists construe the natural world. Biotechnologists can now dissect genetic information taken from one cell and put it into another. The biochemical machinery of the receiver cell decodes the extra information and reproduces the functions of the original cell. Knowing how to interpret the blueprint of life allows biotechnologists to draw up plans for the design of organisms with new functions and properties.

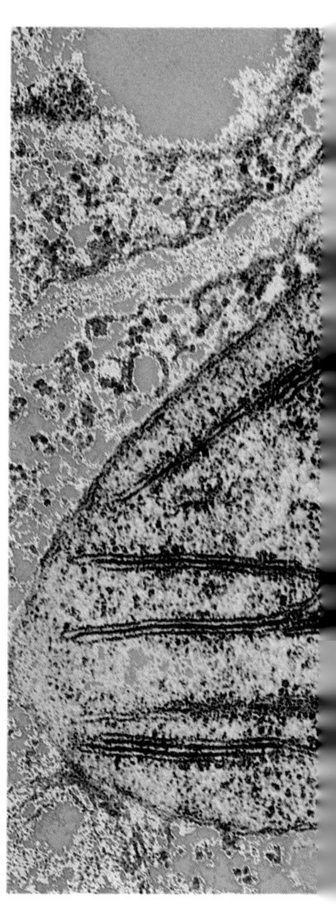

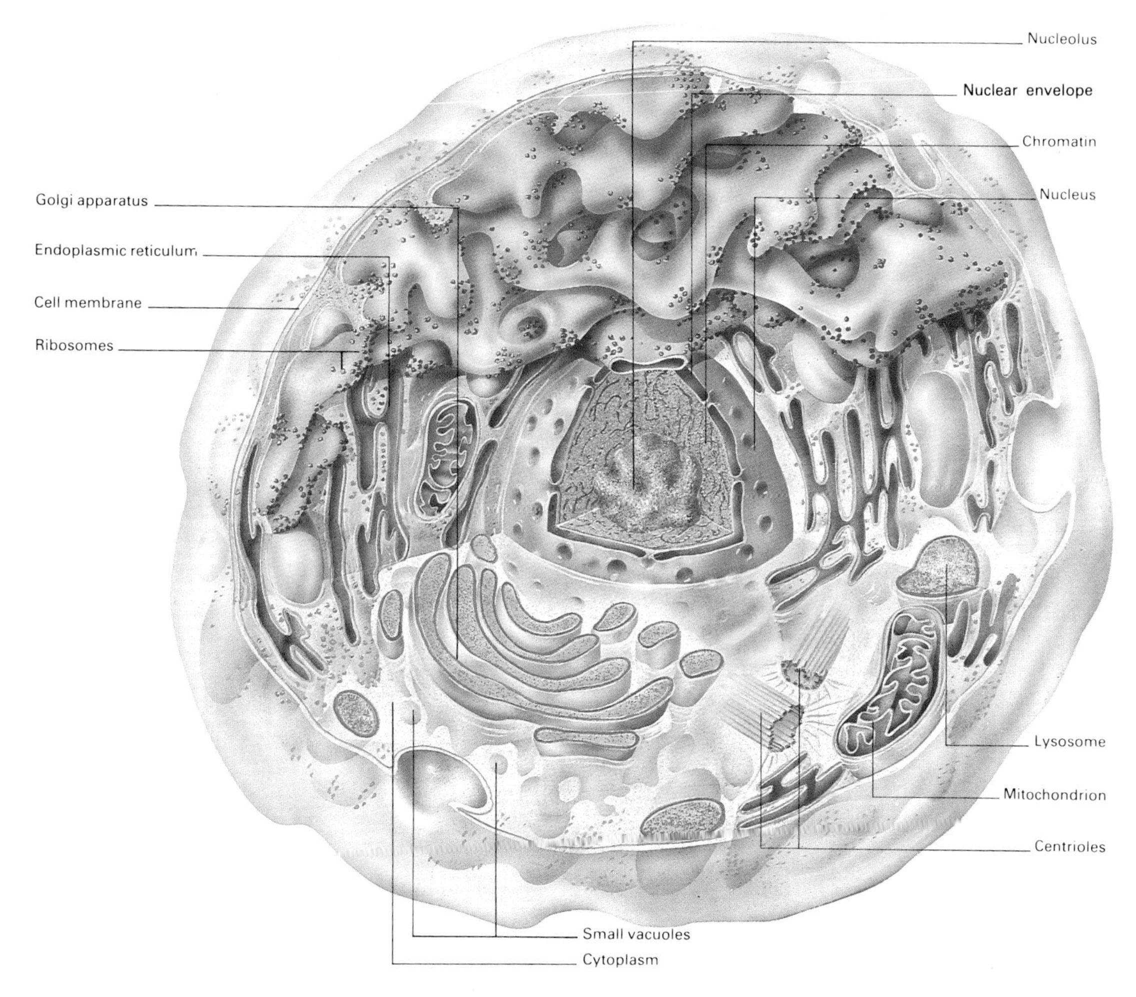

◀ **_A typical animal cell. The nucleus containing the genetic information is surrounded by a membranous envelope. Other organelles, too, have membranes. These allow different biochemical reactions to occur at distinct parts of the cell. Plant cells are similar in their organization but are surrounded by rigid cell walls composed mainly of cellulose._**

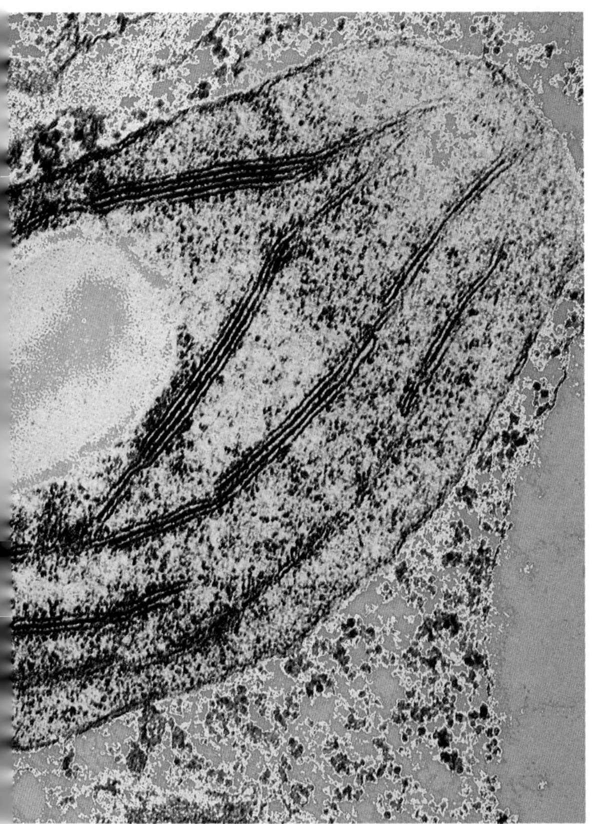

◀ **_A chloroplast from a leaf of a pea plant. Chloroplasts, which may have descended from an ancient blue-green alga, are the organelles responsible for photosynthesis. They convert light and carbon dioxide into chemical energy. Chlorophyll, the light-gathering green pigment, is present in the layered structures of the chloroplast. The central oval shape is a grain of starch, a plant energy store._**

Cellular symbiosis

Primitive eukaryotic cells were probably not very efficient because they could not generate much energy from the biochemicals they used as food. However, this changed very rapidly when the cells were invaded by simple organisms that could produce much larger amounts of energy. In return for this energy, the eukaryotic cell provided protection and food for the invader. This is an example of symbiosis, a mutually beneficial interaction between two organisms.

The first organelle believed by evolutionists to have originated as a symbiotic bacterium is the mitochondrion. Mitochondria are similar in size to bacteria and act as the "powerhouses" of eukaryotic cells. Mitochondrial replication and protein production also resemble those of bacteria. Each mitochondrion contains a small piece of its own DNA – evidence of its once independent existence – which codes for functions relevant to energy generation.

The second important organelle to have descended from a symbiotic prokaryote – the chloroplast – is only found in plants and green algae. By symbiosis the cells gained access to a different source of energy – light energy from the Sun. Like mitochondria, chloroplasts contain their own small piece of DNA which codes for several important functions of the organelle. There may be several hundred chloroplasts in actively photosynthesizing cells.

▲ Gregor Mendel, the father of genetics. An amateur scientist, Mendel showed in 1866 how the characteristics of pea plants pass from one generation to another. His published work went unnoticed by other scientists for over 30 years.

Characteristics and factors

What determines the properties of cells? How do organisms pass on characteristics to their offspring? How do species evolve? Why do many organisms have male and female types? These are the kinds of questions a geneticist attempts to answer.

The science of genetics was born with the experiments of an Austrian monk, Gregor Mendel (1822–1884). Before Mendel's work was published in 1866, and indeed for a long time afterward, scientists believed that the characteristics of parents were blended in their offspring, just as paints are mixed. Mendel disproved this theory with a series of experiments on pea plants grown in the monastery gardens. In explaining his experiments, Mendel made two assumptions which laid the foundation of the science of genetics. First, he supposed that an organism contains unknown "factors" which determine the properties of the organism and those of its progeny. Mendel suggested that these "factors" existed in complementary pairs since each new organism would receive one from each parent. A parental plant would pass on to its offspring only one of each pair of "factors". Second, he postulated that if the paired "factors" differed, one would dominate the other in determining the characteristic of the organism.

Mendel rationalized his results, starting with pure-breeding plants – plants which, when crossed with a plant of the same type, will produce progeny of the same type. Pure-breeding plants therefore contain two "factors" that are the same. In a cross between a pure-breeding short plant and a pure-breeding tall plant, each of the progeny inherit a tallness factor from the tall parent and one for shortness from the short parent. Since tallness dominates, all the progeny are tall even though they contain shortness factors. When these "mixed-factor" plants were interbred, one quarter of the next generation inherited two tallness "factors", another quarter received two shortness "factors", and the remainder inherited a mixture. Only the plants with no tallness "factor" were short. The rest were tall, giving a 3:1 ratio of tall to short plants.

When a second "factor" was introduced into the cross, Mendel observed that the 3:1 ratio for that characteristic was superimposed on the first ratio. A cross between pure-breeding tall plants with smooth peas and pure-breeding short plants with wrinkled peas produced, in the first generation, only tall plants with smooth peas. In the next generation, he saw that the ratio of tall smooth to tall wrinkled to short smooth to short wrinkled peas was 9:3:3:1. Mendel showed therefore that characteristics are inherited according to a rational and predictable pattern. Parents pass on "factors" to their offspring, but the characteristics of the progeny depend on the way the inherited factors interact. The term "gene" was later adopted for Mendel's "factors" determining inherited characteristics.

In 1868, two years after Mendel's work on pea plants, a Swiss biochemist, Friedrich Miescher (1844–1895), identified a polymer (complex substance composed of many identical or similar subunits) containing phosphates in pus cells on discarded surgical bandages. This was deoxyribonucleic acid, or DNA. Miescher and other investigators subsequently found it in cells from a wide range of organisms. At that time no connection was made between DNA and heredity. Indeed, DNA was largely ignored as a possible genetic material even after Mendelian principles of genetics became established and accepted among the scientific community.

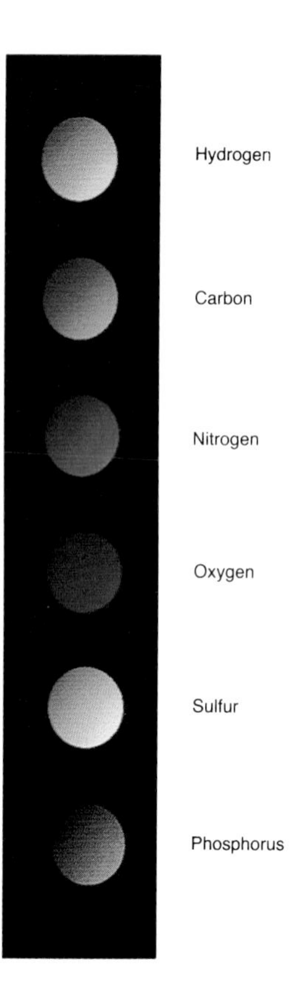

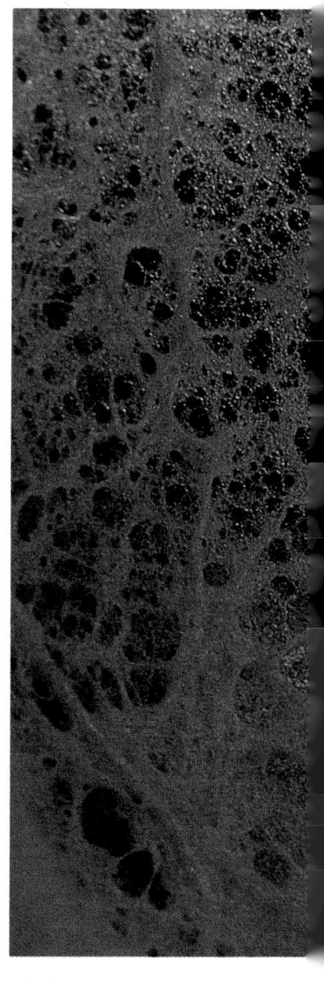

Chromosomes, genes and DNA

In the latter years of the 19th century, the chromosome was discovered. Chromosomes, seen under a microscope, are dark-staining strands in the nucleus of the cell. The chromosomes occurred in pairs and were highly organized so that when cells divided, each daughter cell received a complete set of chromosomes. In 1903, William S. Sutton (1877–1916), an American biologist familiar with Mendel's experiments on peas, suggested that the chromosomes were the physical basis of heredity.

By 1900 it was known that chromosomes contained both DNA and protein. Chemists had broken DNA down into its fundamental components and analyzed them. The composition of DNA was simple and appeared to be very similar in all organisms. On the other hand, proteins were highly complex molecules made up of 20 different subunits. The prevailing opinion, therefore, was that proteins, not DNA, were the genetic material. DNA was regarded merely as a "midwife" molecule which held proteins in a linear conformation while protein duplication took place.

In the early 1940s, the American chemist Oswald Avery (1877–1955) produced a 99.98 percent pure preparation of DNA (0.02 percent protein), which he used to transform bacteria. Its transforming activity was destroyed by the enzyme DNAase, which destroys DNA, but not by enzymes which only break down protein. This finally convinced most people that it was DNA, after all, that was the "transforming principle", the elusive genetic material.

◀ Biological matter is mainly made up of six types of atoms (shown as colored spheres). Groups of atoms joined together are called molecules. The links between atoms are known as covalent bonds and are shown as yellow sticks in "ball-and-stick" models. In "space filling" models the atoms are to scale and the bonds are not shown.

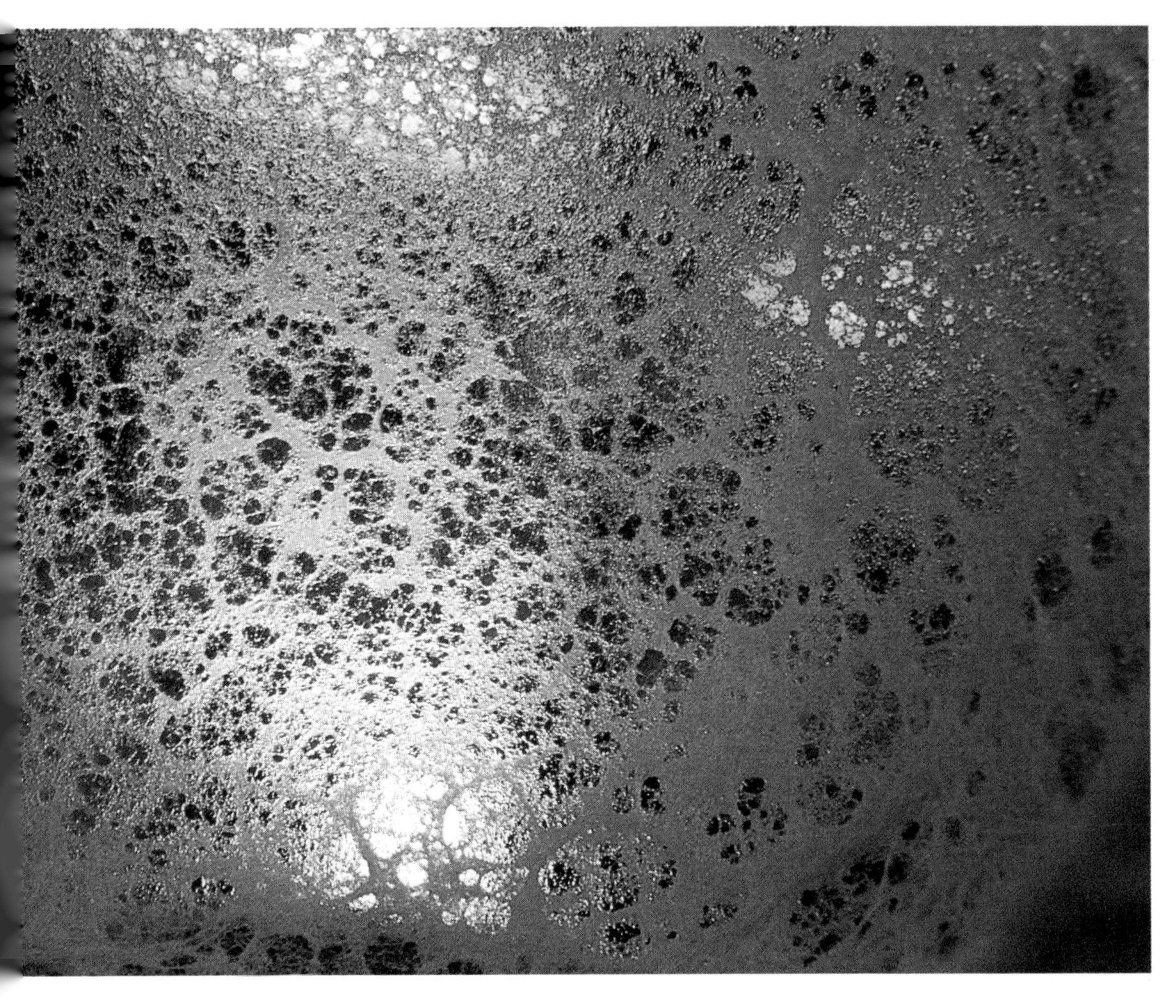

◀ A specimen of pure DNA, the genetic material. Scientists obtain such a preparation by finely mincing up biological tissues and treating them in such a way as to disrupt all the cells. They then use a centrifuge to remove the debris of broken cell walls and membranes. The next problem is to remove all the other large molecules (protein, RNA) from the soup that remains.

The Amino Acids

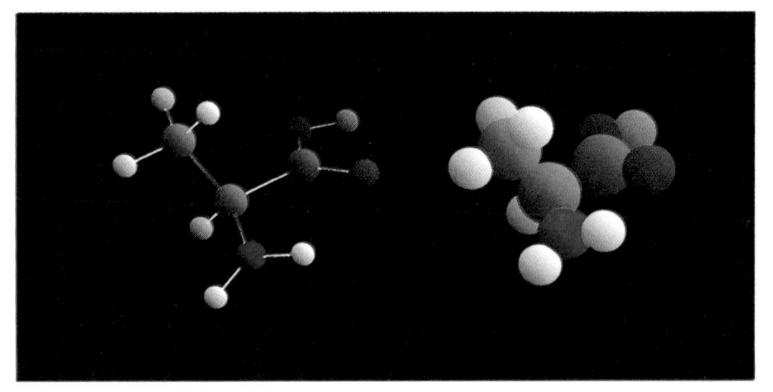
Alanine

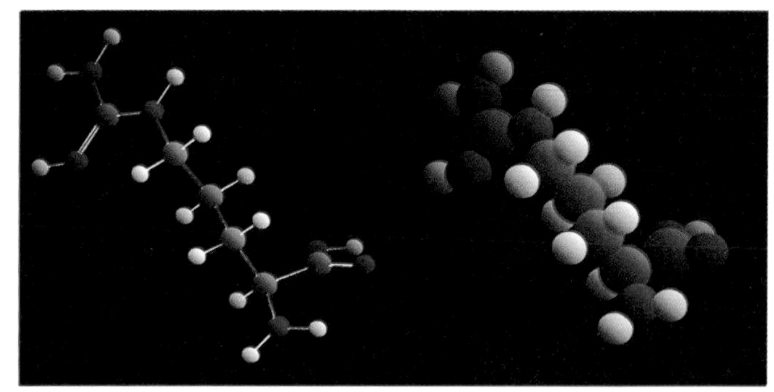
Arginine

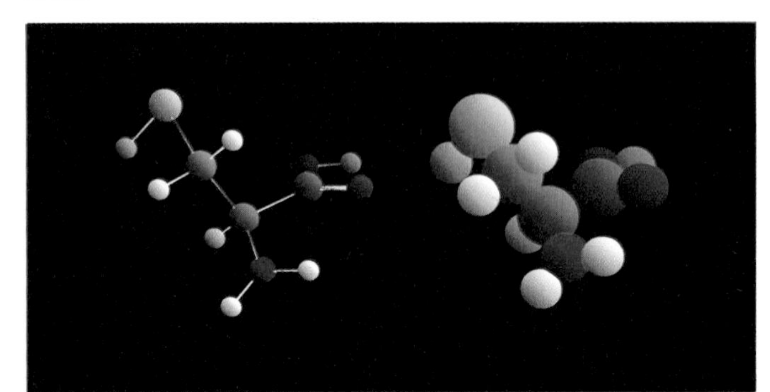
Cysteine

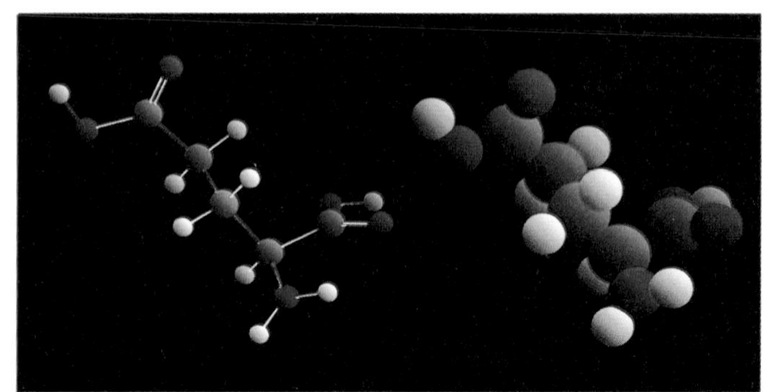
Glutamic acid

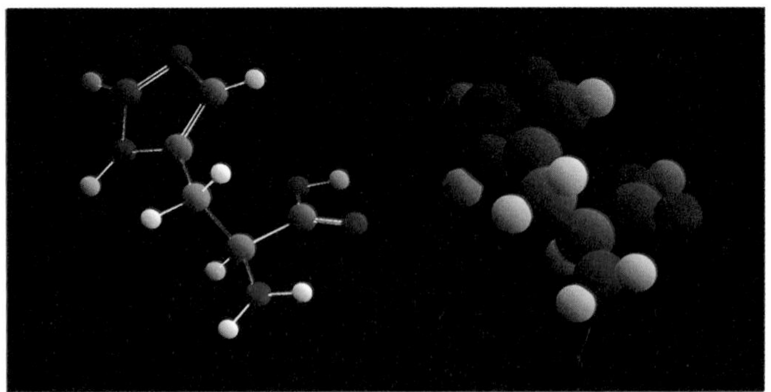
Histidine

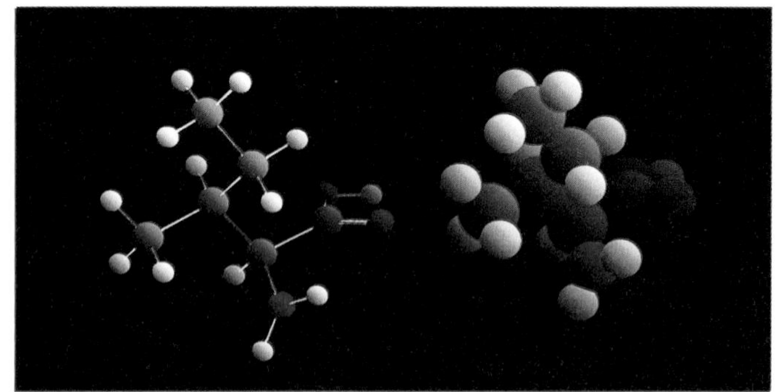
Isoleucine

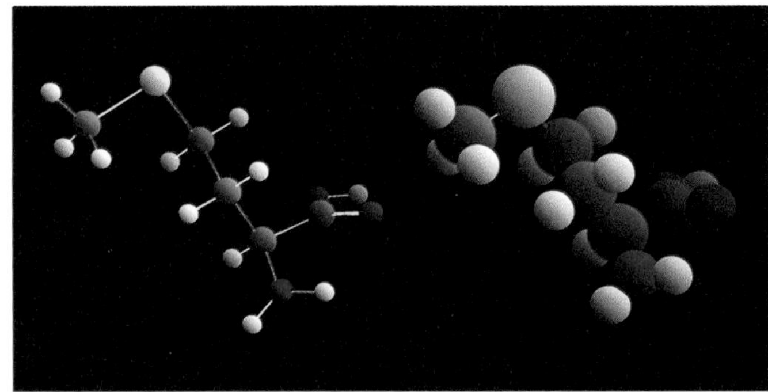
Methionine

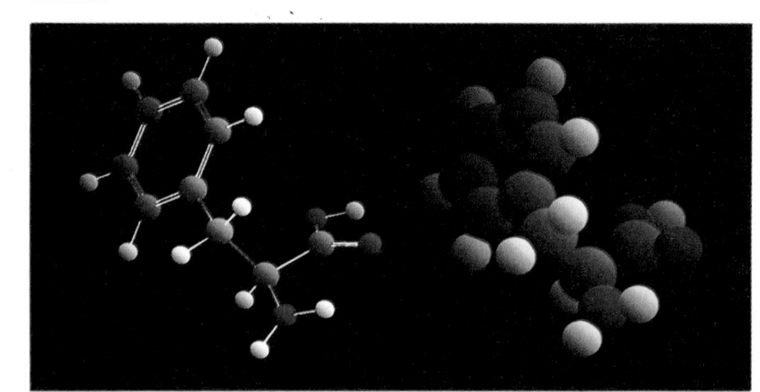
Phenylalanine

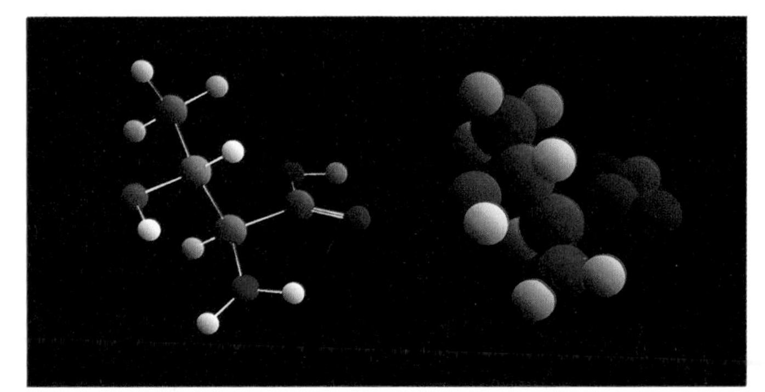
Threonine

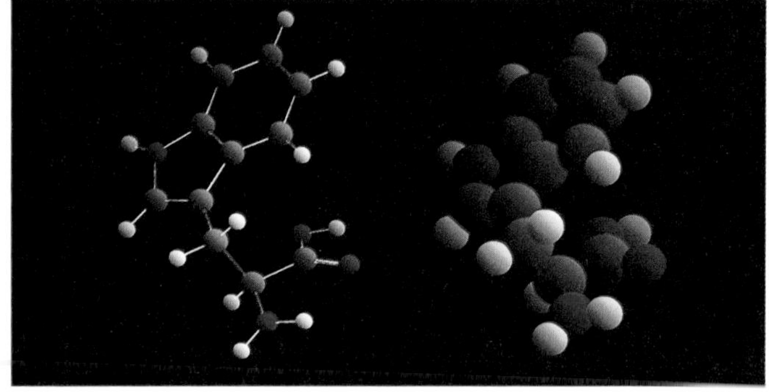
Tryptophan

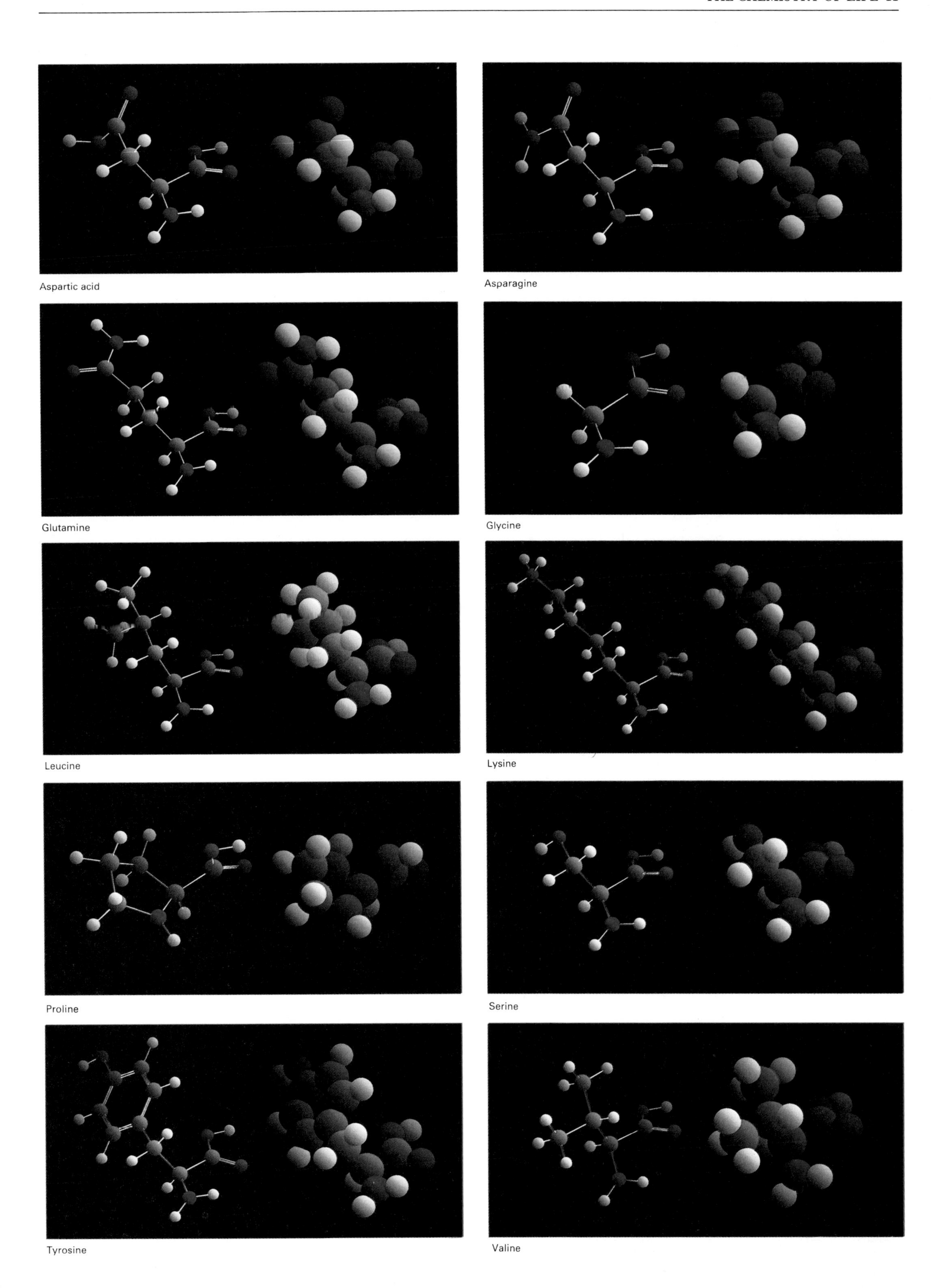

Aspartic acid

Asparagine

Glutamine

Glycine

Leucine

Lysine

Proline

Serine

Tyrosine

Valine

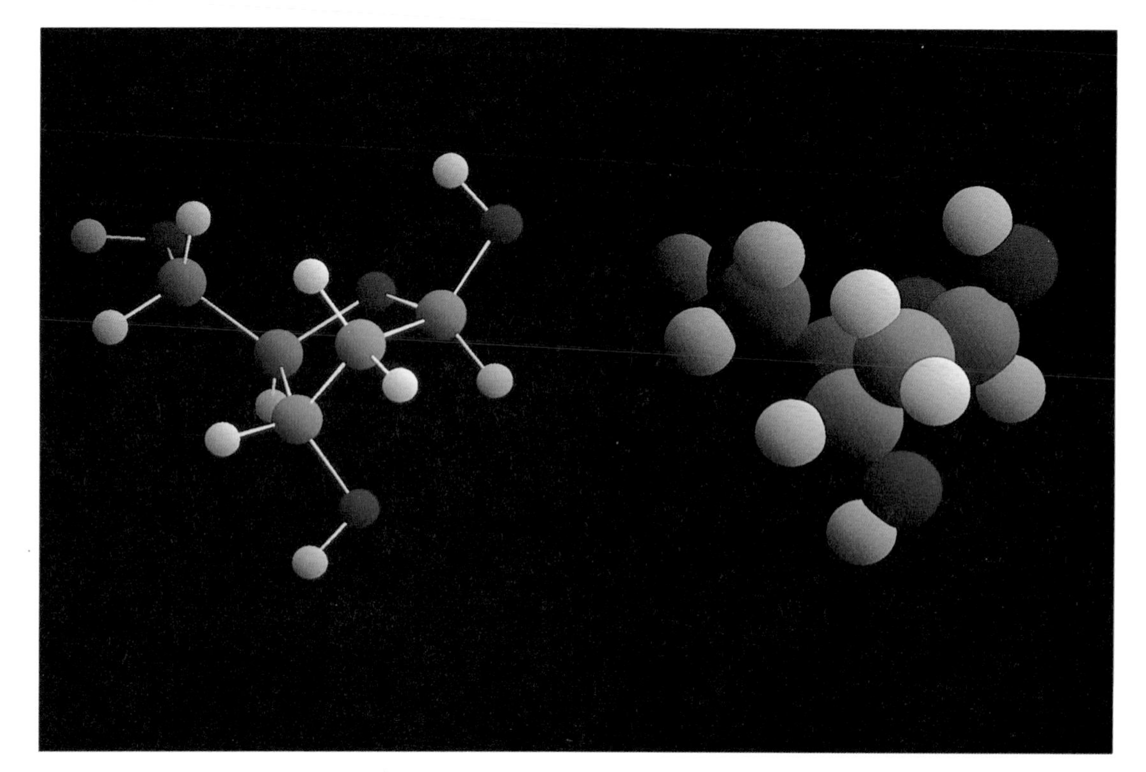

► ***The sugar molecule deoxyribose. Its three reactive oxygen atoms (those attached to hydrogen) are all involved in linking together the various chemical subunits of DNA. The absence of a fourth reactive oxygen atom (hence deoxy-) makes DNA chemically very stable.***

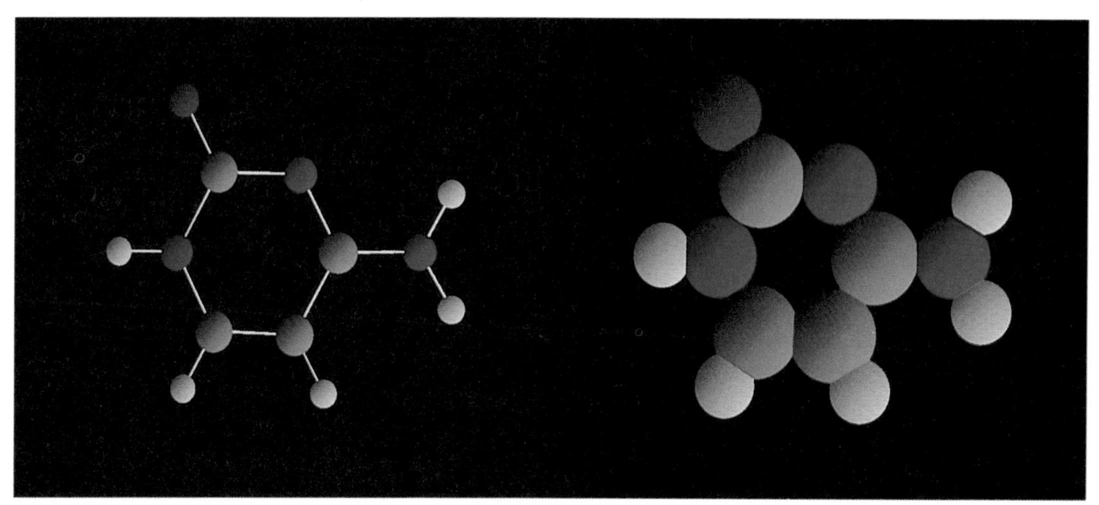

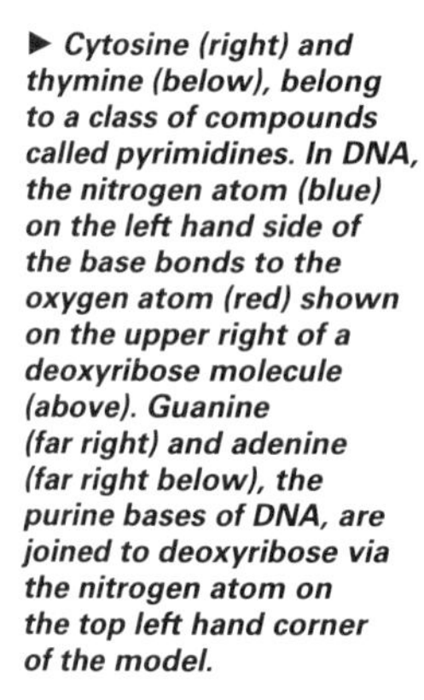

► ***Cytosine (right) and thymine (below), belong to a class of compounds called pyrimidines. In DNA, the nitrogen atom (blue) on the left hand side of the base bonds to the oxygen atom (red) shown on the upper right of a deoxyribose molecule (above). Guanine (far right) and adenine (far right below), the purine bases of DNA, are joined to deoxyribose via the nitrogen atom on the top left hand corner of the model.***

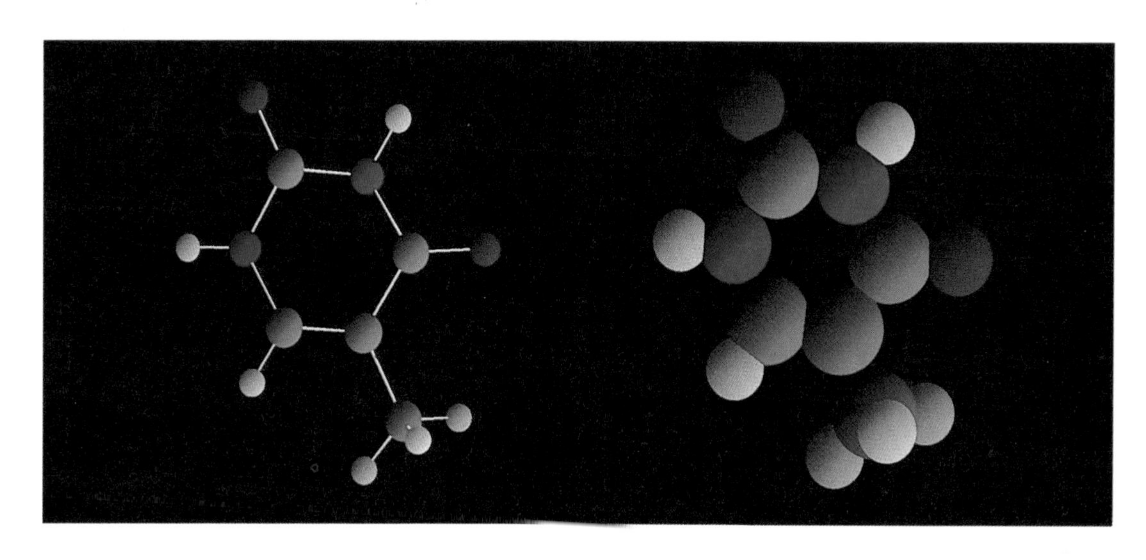

The composition and structure of DNA

Since the beginning of the 20th century, chemists had been trying to work out the chemical structure of the large molecules found in the nucleus of the cell, the "nucleic acids". They knew that these substances always contained phosphate (an acidic compound of phosphorus and oxygen) molecules of a particular sort of sugar, and several different kinds of organic compounds called bases. The nucleic acid found in the chromosome was known as deoxyribonucleic acid (DNA) since chemical analysis had shown that it had a high content of the sugar, deoxyribose. There were four different bases in DNA, each containing carbon and nitrogen atoms arranged in rings. These were adenine (A), guanine (G), cytosine (C) and thymine (T). However, chemists could not guess how these subunits might be assembled.

At first, they thought the bases occurred in equal amounts in the DNA of all organisms, a belief which could be held as evidence against DNA being the genetic material. However, Erwin Chargaff (b.1905) and his colleagues in the late 1940s and early 1950s found that the base composition of DNA varied considerably from species to species. In humans there is 1.5 times as much adenosine (A) and thymidine (T) as there is cytosine (C) and guanine (G). In the fungus *Aspergillus niger* there is an approximately equal amount of each of the bases. In other organisms, there is much less A and T than G and C. Thus there was sufficient variation for DNA to be the genetic material.

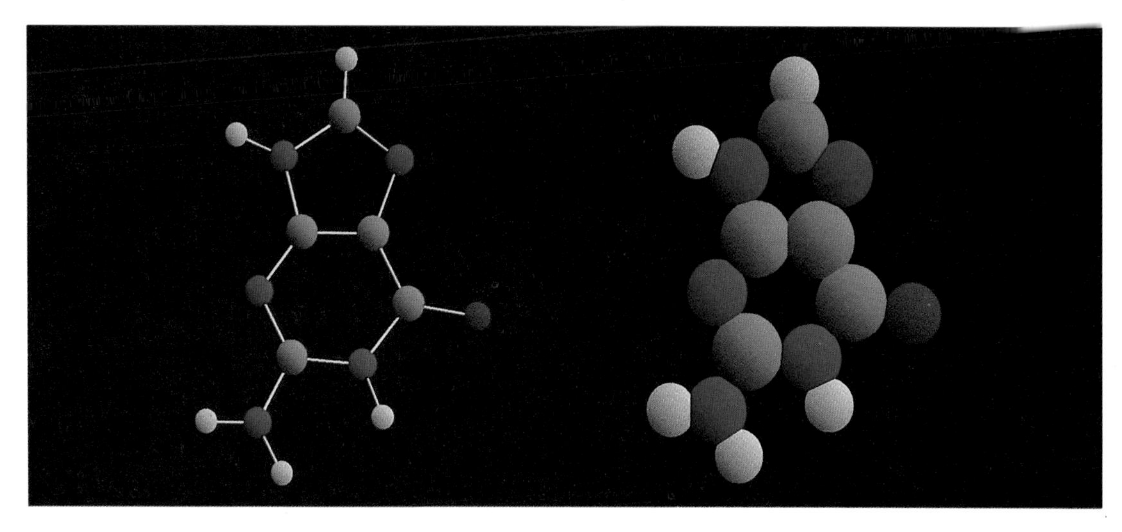

◀ *Erwin Chargaff found that the amounts of the four DNA bases vary from organism to organism. Moreover, the amount of adenine is always equal to the amount of thymine, and cytosine always equals guanine. Also, the amount of purines (G and A) equals the amount of pyrimidines (C and T). Chargaff's observations were clues that led to the elucidation of the structure of DNA.*

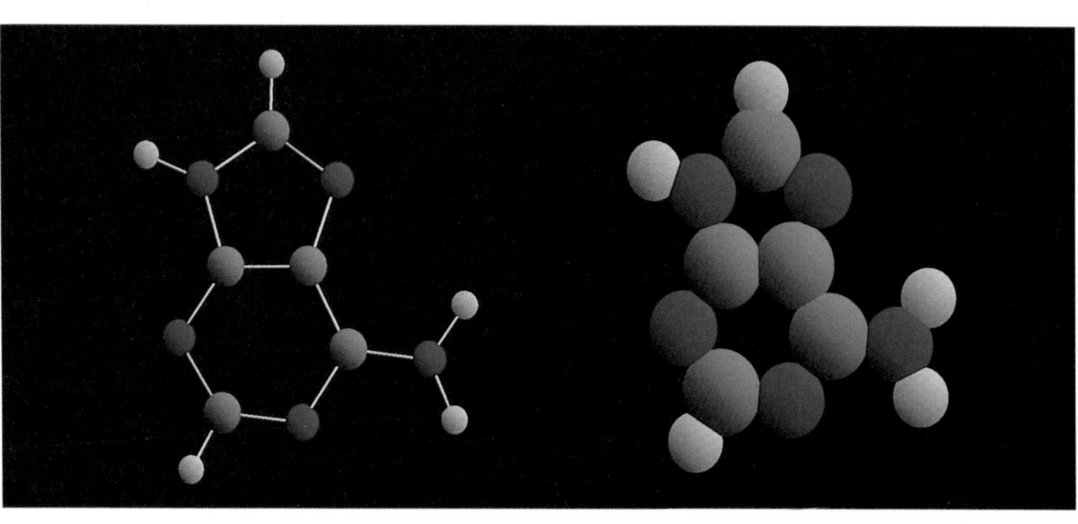

While Chargaff was analyzing the composition of DNA, the English chemist, Alexander Todd (b.1907) was investigating its structure. He showed that each sugar molecule was connected, through phosphate groups, to two other sugars. DNA, he concluded, was a long, linear molecule with a backbone of alternate sugar (deoxyribose) and phosphate molecules. Each sugar also had a single base attached to it. DNA is a polymer (Greek for "many parts") and the repeating sugar–phosphate-base units are called nucleotides.

In the early 1950s, at King's College in London, Rosalind Franklin (1920–1958), a chemist, and the physicist Maurice Wilkins (b.1916) examined the structure of DNA using X-ray crystallography. Their results allowed James Watson (b.1928) and Francis Crick (b.1916) to postulate a structure for DNA in 1953. Crick and Watson proposed that DNA is a helix in which two strands of DNA run parallel but in opposite directions. In the center of the molecule, weak chemical interactions called hydrogen bonds hold the two strands together. The bonds form between bases. Adenosine always forms hydrogen bonds with thymine, cytosine always with guanine. Other combinations (A=G, A=C, G=T, C=T) are improbable because the base pairs cannot form appropriate hydrogen bonds. The sugar–phosphate backbones of the two strands are on the outside of the double helix.

The Watson–Crick double helix-model of DNA was a triumph. It drew upon and accounted for all the known physical and chemical properties of DNA. Moreover, the pairing of the bases between strands provided a mechanism for gene duplication. If the double helix unzipped, each strand could serve as a template to direct the synthesis of a new complementary strand. Scientists now knew how one DNA molecule could become two. They had discovered how DNA replicated.

The sequence of the bases attached to the sugar–phosphate backbone in DNA is the information of life. Somehow long chains of DNA containing only four different subunits (the bases A, T, C and G) direct the production of proteins, which are long chains containing 20 different building blocks (amino acids). The information of DNA has to be converted into the information of proteins. Many plausible genetic codes could be envisaged.

►► The base pairs of DNA. Watson and Crick used cardboard shapes to work out how the DNA building blocks fitted together. As the computer models show, the four bases make two pairs – adenine bonds with thymine, and guanine with cytosine (far right). Weak forces called hydrogen bonds (the longer bonds) hold the pairs together. These can form when a positively charged hydrogen atom (white) is sandwiched between two negatively charged nitrogen (blue) or oxygen (red) atoms. Their three hydrogen bonds make GC pairs stronger than AT pairs.

► The base pairs (balls) in the center of the DNA spiral hold the two strands together. To duplicate DNA, the hydrogen bonds break and the double helix unzips. The resultant single strands are templates for two new double helices. The specific bonding between bases dictates that both new helices are identical to the old one.

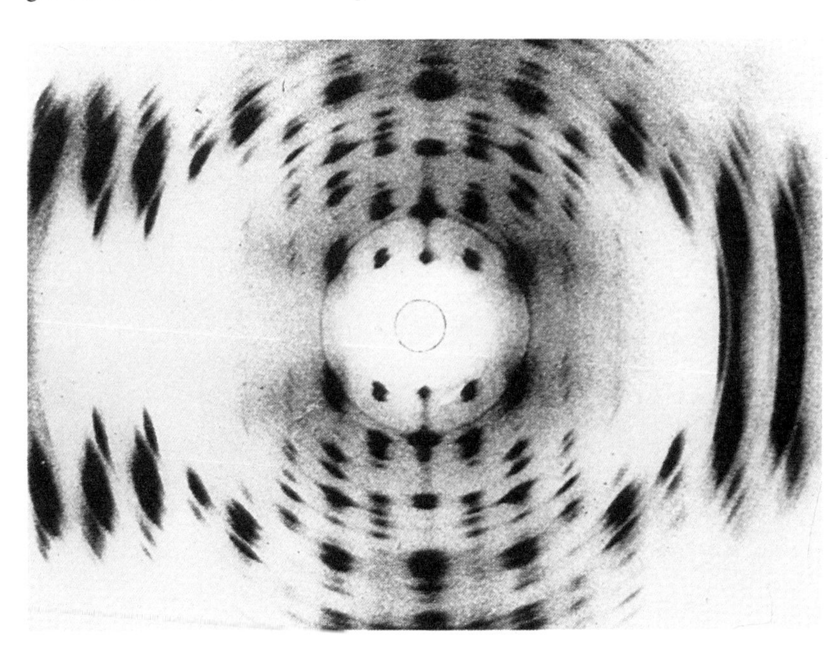

► An X-ray diffraction image of DNA of the type seen by Watson and Crick during a visit to the laboratories of Rosalind Franklin and Maurice Wilkins at King's College, London. The pattern and spacing of the dots gave them vital information about the structure of DNA, suggesting – most importantly – that it was shaped like a coiled spring or helix.

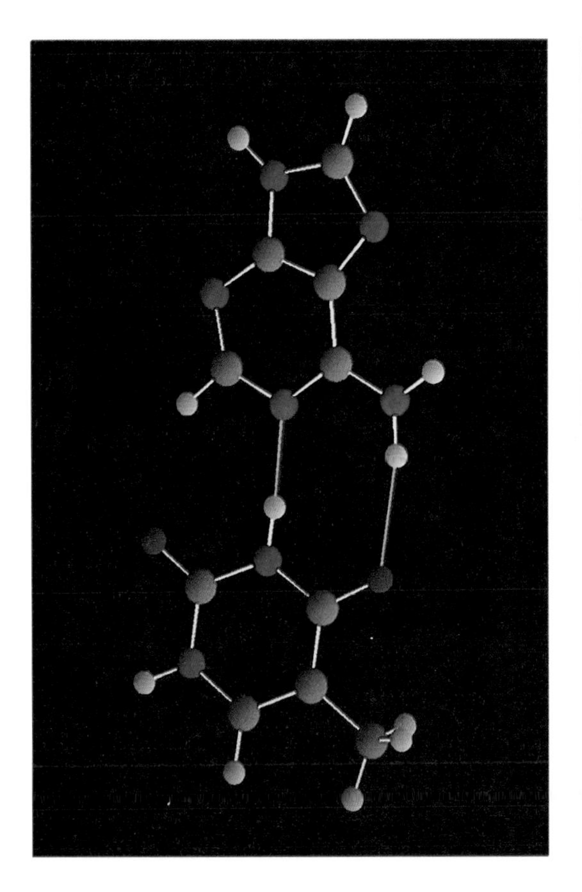

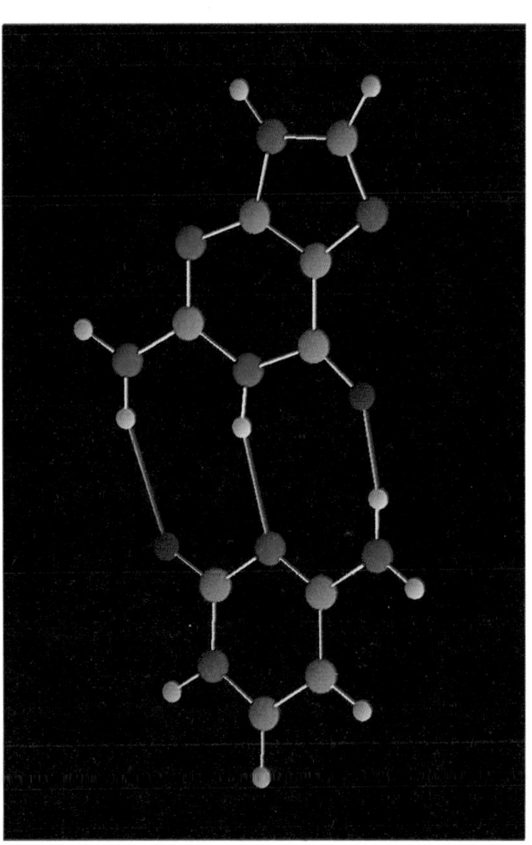

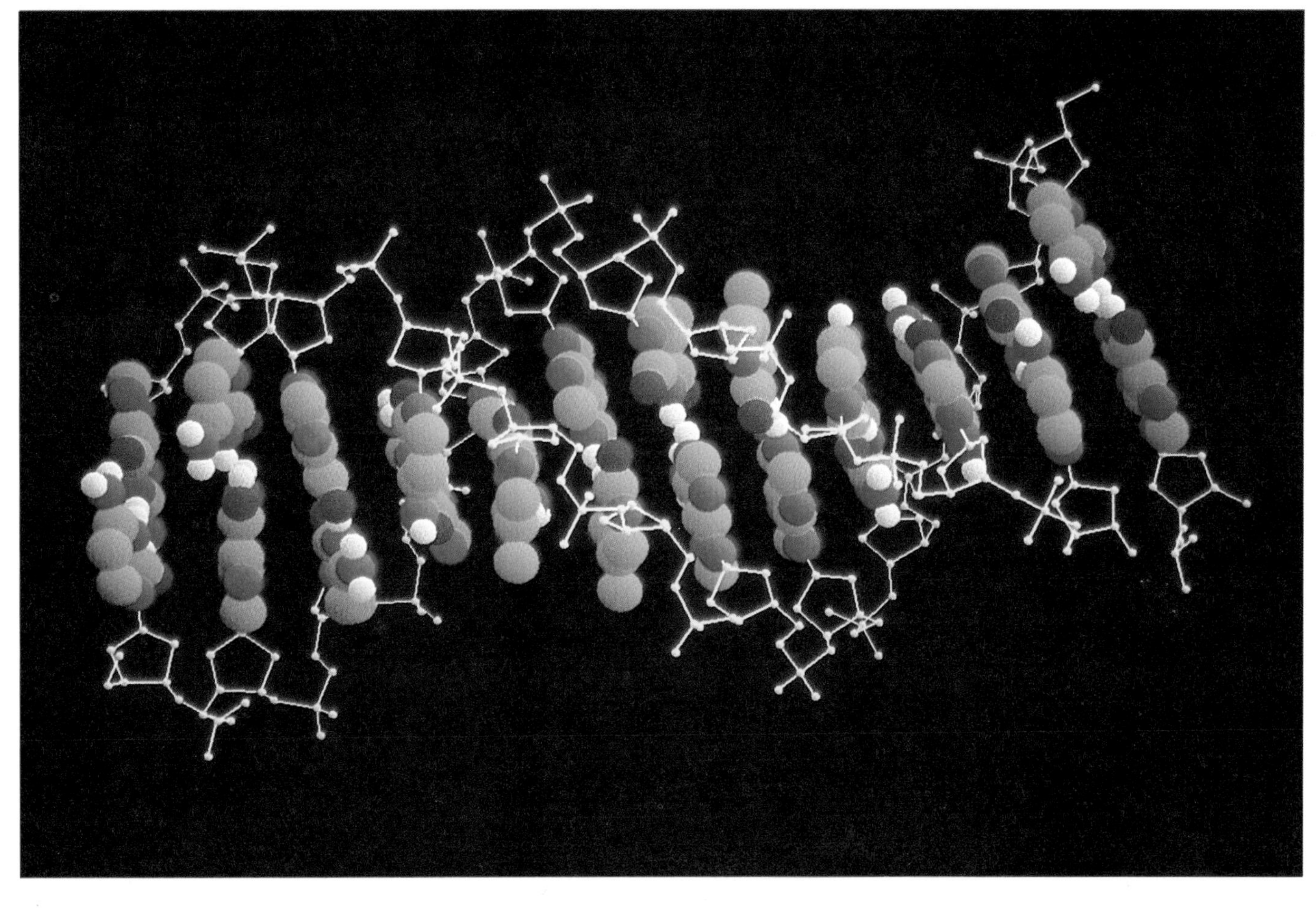

The DNA Backbone

► *The backbones of DNA are on the outside of the molecule. They do not stiffen the molecule like an animal's vertebrae, but provide regular sites of attachment for the bases on the deoxyribose molecules. The two strands of DNA are "antiparallel", that is, the backbone of one runs in the opposite direction to the backbone of the other.*

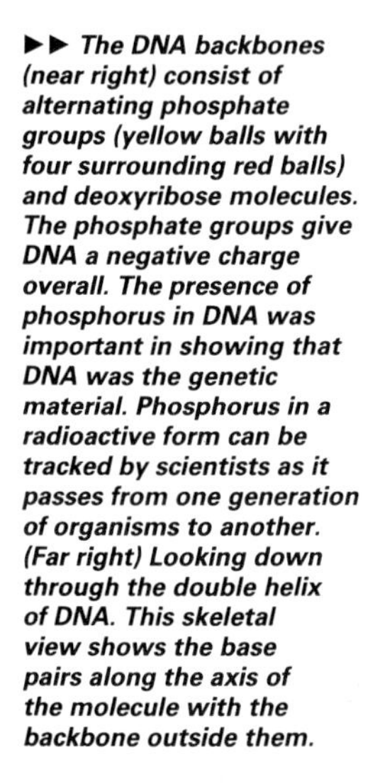

►► *The DNA backbones (near right) consist of alternating phosphate groups (yellow balls with four surrounding red balls) and deoxyribose molecules. The phosphate groups give DNA a negative charge overall. The presence of phosphorus in DNA was important in showing that DNA was the genetic material. Phosphorus in a radioactive form can be tracked by scientists as it passes from one generation of organisms to another. (Far right) Looking down through the double helix of DNA. This skeletal view shows the base pairs along the axis of the molecule with the backbone outside them.*

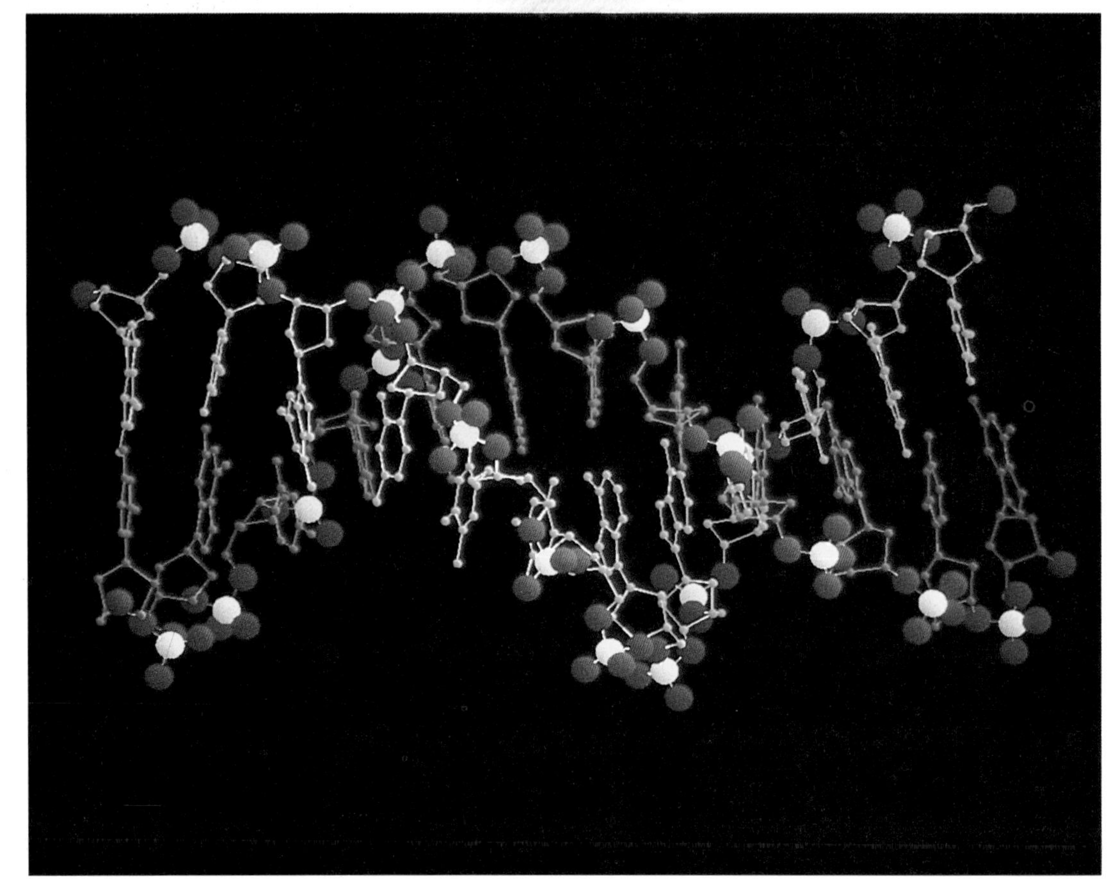

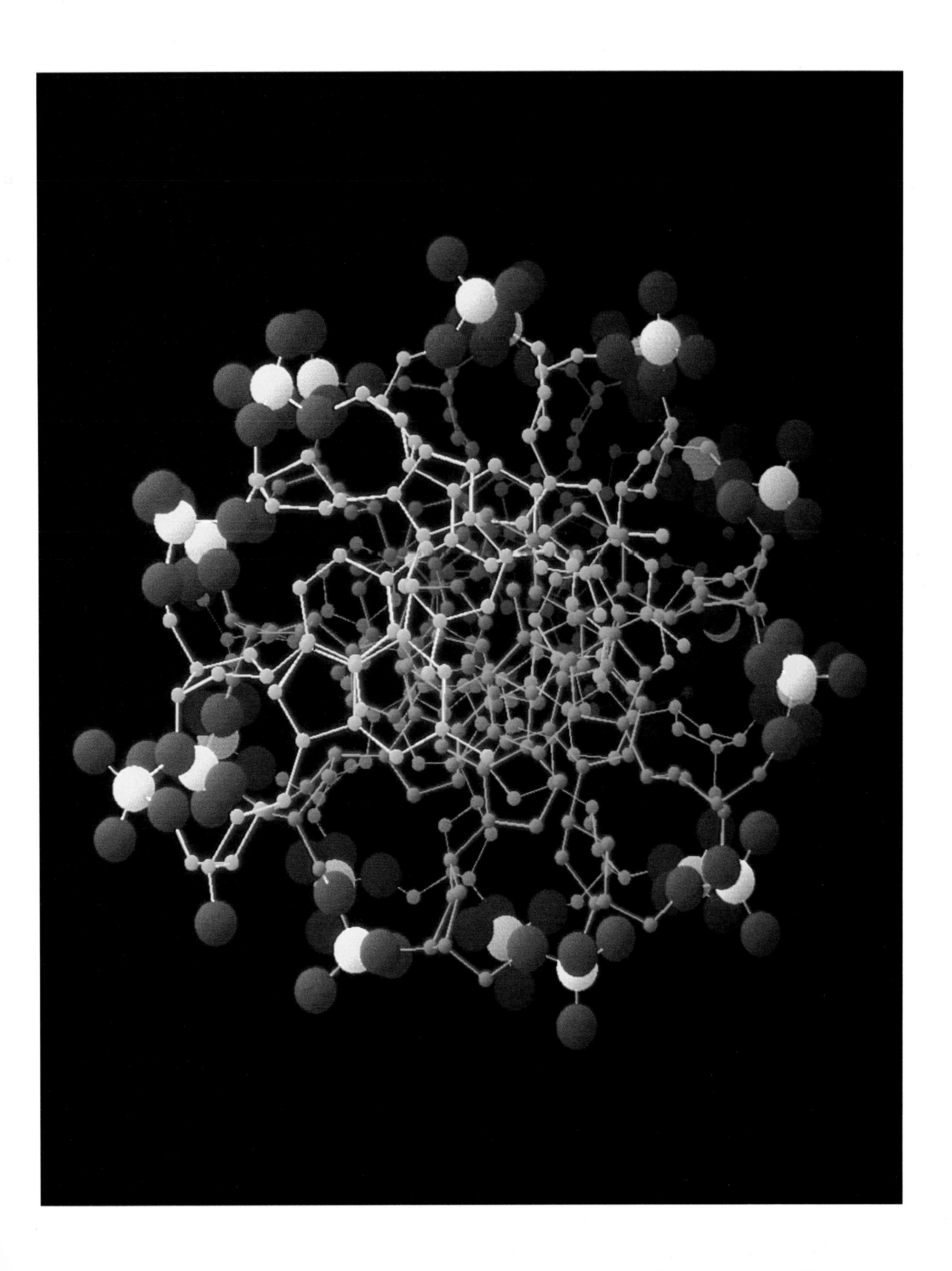

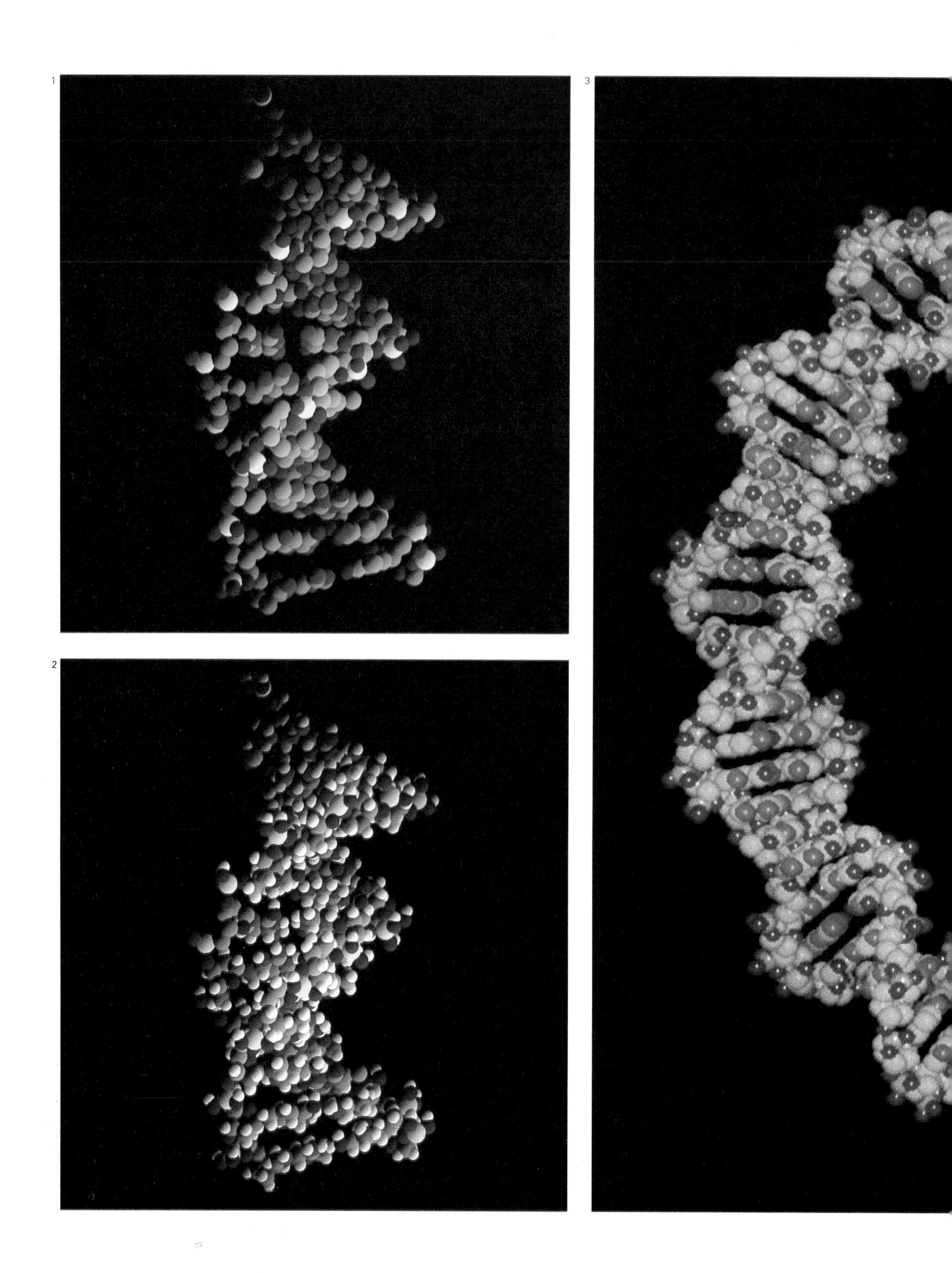
1
2
3

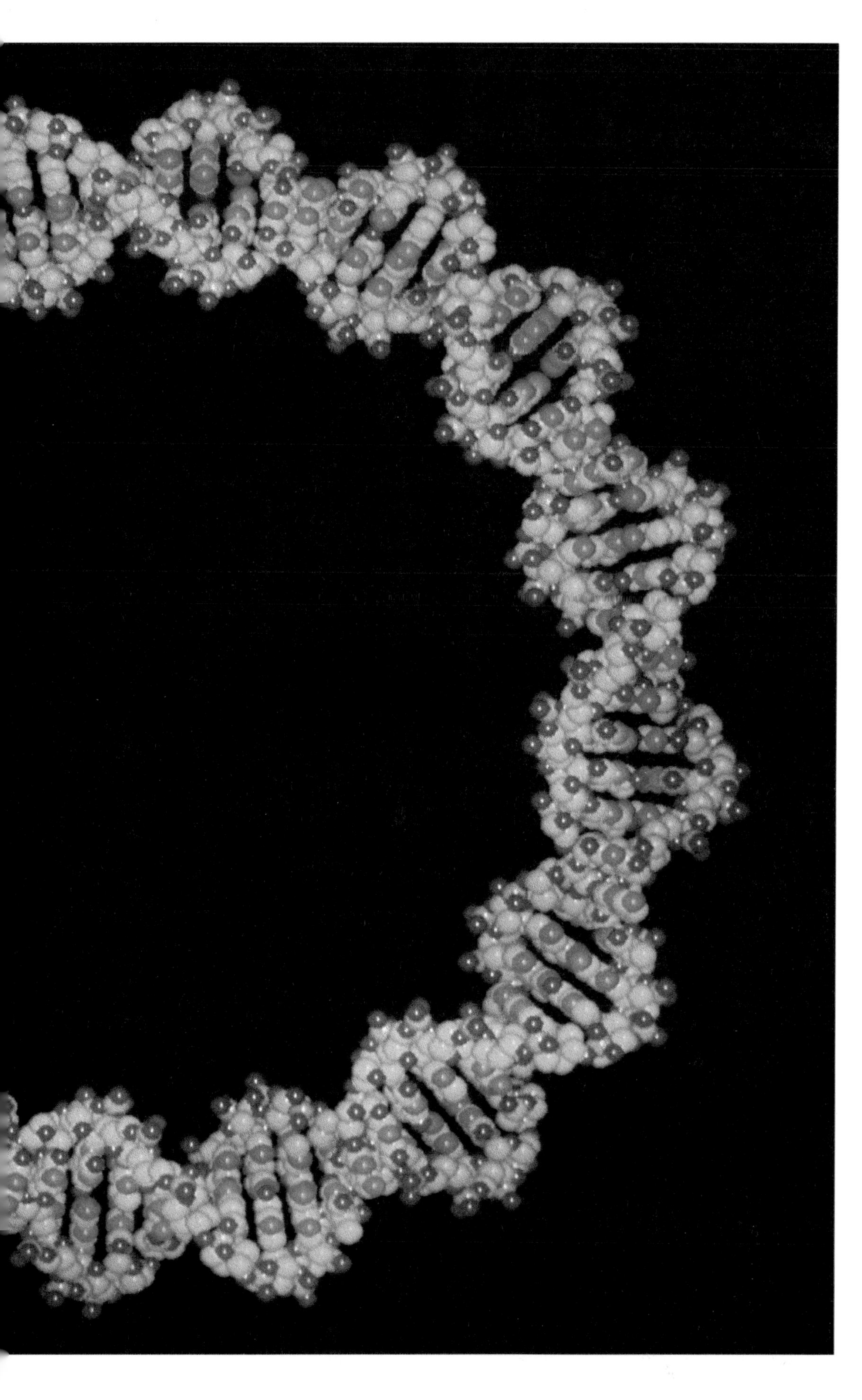

The backbones of DNA's two strands wind round in a helix (1). A minor and a major groove track along the length of the helix. Each full turn of the helix contains base pairs and is 3.4 nanometers (billionths of a meter) in length. DNA is an immensely complex molecule composed of many individual atoms. Each turn of the double helix contains some 700 atoms of hydrogen, carbon, nitrogen, oxygen or phosphorus, the precise arrangement of which is different in every type of organism (2). The circle of DNA which has about 11,000 atoms represents only one twentieth of one millionth of the amount of DNA present in every human cell (3).

Breaking the code

In 1961, Francis Crick and Sydney Brenner (b.1927) working in Cambridge, UK, demonstrated that each amino acid in a protein is coded for by a set of three bases of DNA, which they called codons. They treated DNA from a virus with chemicals known as acridines. These caused mutations by deleting base pairs from the DNA sequence. Normally the virus DNA codes for a functional protein. When Brenner and Crick deleted one or two base pairs from the DNA, the protein for which it coded became completely useless. If they deleted three base pairs, however, the protein behaved virtually as normal. This showed that a DNA sequence has to be read in groups of three bases. Moreover, deleting base pairs at one end of the sequence had much less effect on the activity of the protein than base-pair deletions at the other. The DNA code had directionality.

The DNA codon is a triplet of bases. But which triplet specifies which amino acid? Biochemists had been able to "tag" amino acids with radioactive labels to trace which of the acids became incorporated into protein. In effect, they could read the decoded version of DNA. Furthermore, they already had a decoding machine, the cell, which converted the cryptic message (DNA sequences) into protein. What they lacked were the original cryptic messages. To discover these, they produced synthetic sequences of RNA (ribonucleic acid), a close relation of DNA. A particular type of RNA, messenger RNA (mRNA), was already known to be involved in the decoding of DNA. They then fed synthetic mRNA to the cell's biochemical decoding

▼ In sequences of RNA, uracil occurs in place of the thymine base in DNA. The two bases are closely related. Thymine is also known as methyl-uracil because it has an extra CH_3 group. The scientists who worked out the genetic code synthesized RNA rather than DNA as input to the cell's decoding machinery. Thus the code usually refers to sequences containing uracil rather than thymine.

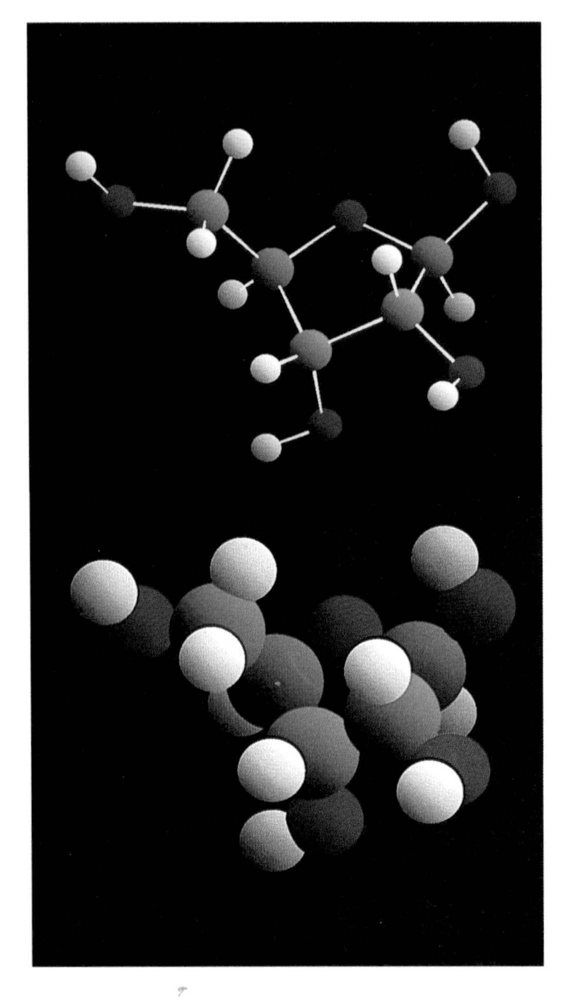

► Ribose, the sugar found in RNA, is identical to deoxyribose except that it has an extra oxygen atom (bottom right of molecule), which makes it more reactive. Unlike DNA, RNA usually occurs in a single-stranded form so that it can adopt a variety of shapes including some that allow it to promote biochemical reactions. The earliest organisms may have contained RNA which could reproduce itself.

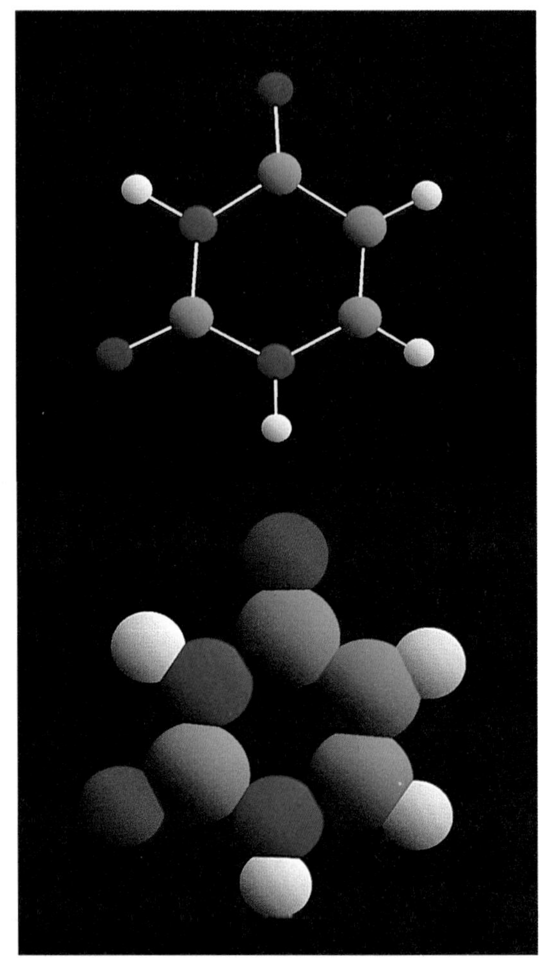

machinery and analyzed which amino acids appeared in the resulting protein. The first synthetic RNA to be fed was a polymer containing just uracil (poly-U). Only the amino acid phenylalanine became incorporated into the protein. The codon for phenylalanine is, therefore, a triplet of uracil, UUU. Later experiments using more complex synthetic polymers that contained combinations of A, U, G and C revealed the rest of the genetic code. ACACACACACAC, for example, produced chains of amino acids in which threonine (ACA) was linked to histidine (CAC).

It took from 1961 to 1965 to assign each of 61 triplet codons to one or other of the 20 amino acids that occur in proteins. The other three codons – UAG, UAA and UGA – are not associated with particular amino acids. Rather, they are signals for terminating protein production – the equivalent of stops marking the end of a mRNA sentence. The first amino acid produced in all proteins is methionine (this may be removed subsequently from the protein by the cell if it interferes with the protein's activity). Thus the starting signal for protein synthesis is the methionine codon, AUG.

A sequence of mRNA is interpreted by beginning at AUG and moving along the molecule three bases at a time. Generally, there is only one correct reading frame for a particular sequence of mRNA. Adding bases to or deleting them from genetic sequences will put the reading frame out of phase. Such changes are known as frameshift mutations, and as Crick and Brenner discovered, they cause the production of useless proteins.

▼ The molecular adaptor – transfer RNA (tRNA) – demonstrates how flexible RNA is in the shapes it can adopt and the roles it can play. RNA is a linear molecule comprising a sugar-phosphate backbone (red) to which bases are attached. The backbone weaves its way round itself and the structure of RNA is stabilized by pairing between bases (blue) within the molecule (rather than with a complementary molecule as in DNA). The yellow region is the three-base anticodon which "plugs" into the complementary three-base region on mRNA. The green region is where a protein building block, an amino acid, is bound.

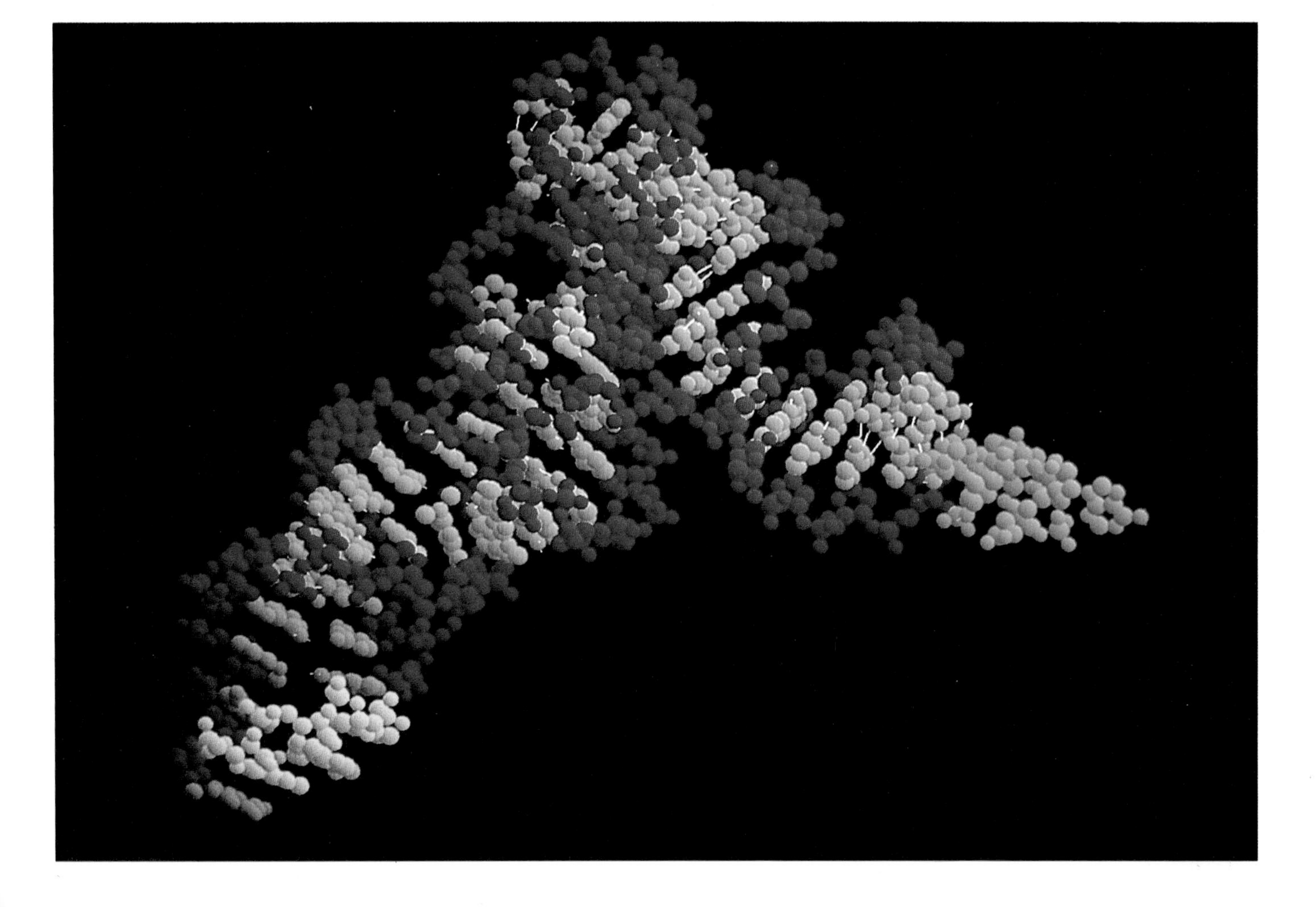

The decoding machine

The replication mechanism is implicit in the DNA structure. But how does DNA, which in eukaryotic cells is retained in the nucleus, direct the production of protein outside the nucleus? The answer lies with three types of ribonucleic acid (RNA) which are components of the cell's decoding machinery – messenger RNA (mRNA), ribosomal RNA (rRNA) and transfer RNA (tRNA).

The nucleus of the cell stores the majority of the cell's genetic information as extremely large molecules of double-stranded DNA. This is packed tightly as complex coils and organized into chromosomes. When instructions are required by the rest of the cell, a small part of the coded messages of DNA is converted into a more mobile form, messenger ribonucleic acid (mRNA). In this process, called transcription, the DNA double helix untwists at particular positions to allow an enzyme called RNA polymerase to bind to the coding strand. RNA polymerase, in effect, examines the first base in the DNA sequence and selects the appropriate RNA unit to go with it, pairing U with A, A with T and C with G. All the RNA building-blocks are present in abundance in the nucleus. The enzyme moves one base along the DNA strand, selects the next RNA unit, and links the two units, before moving on to the third DNA base. This process is repeated until the chain of RNA is complete. Hundreds of chains of RNA can be produced simultaneously from a sequence of DNA, giving a feathered appearance to actively transcribed regions. When completed, the RNA molecules pass out of the nucleus through microscopic pores in the nuclear membrane.

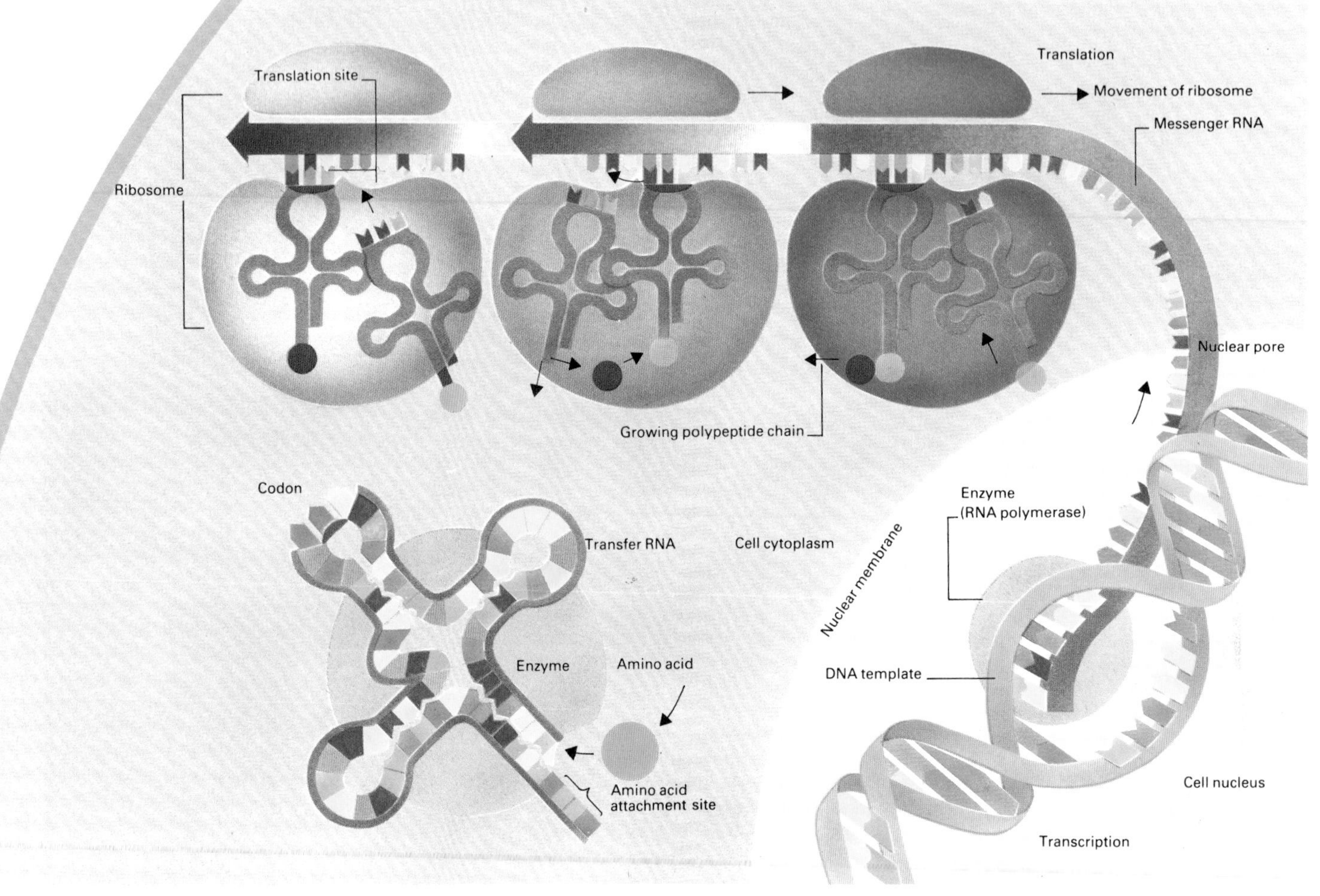

DNA replication

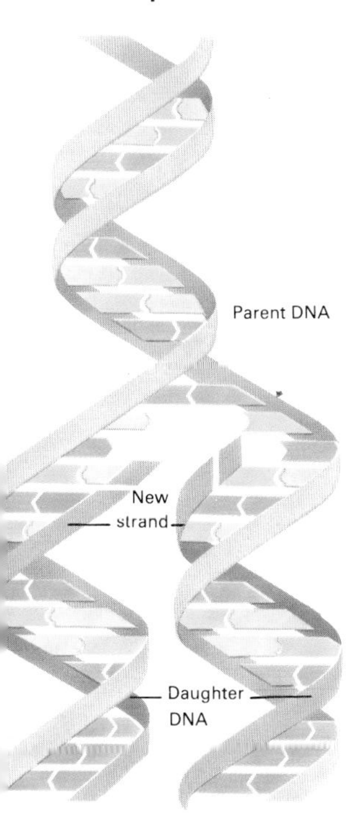

◀ *The production of all proteins starts when the genetic message on DNA is transcribed onto messenger RNA. Only one strand of DNA, the coding strand, is transcribed. Ribosomes latch on to specific regions of mRNA which precede the protein coding sequence. By tracking along the mRNA, the ribosome comes into register with the first and second codons of the gene and attracts transfer RNAs charged with amino acids. The ribosome links the amino acids so that both are attached to one tRNA. The first tRNA, discharged of its amino acid, leaves the ribosome. The ribosome shifts one codon along the mRNA and then a new charged tRNA can add its amino acid to the growing protein chain.*

◀ *DNA replication occurs at a fork where enzymes (DNA polymerases) add DNA building blocks to existing template strands. Enzymes called gyrases must first unwind the DNA from its double helix for this to take place. Other enzymes check that the template strand has been copied correctly.*

From the nucleus, the mRNA moves to the endoplasmic reticulum, another of the cell's organelles. This is a highly convoluted and dynamic membrane which spreads extensively through the cell. It provides a large surface for the acceleration of biochemical reactions. This is where the decoding of mRNA to produce proteins takes place. Biochemists call this decoding "translation". The endoplasmic reticulum acts as a sort of cellular plumbing, a conduit for mRNA, protein building-blocks (amino acids) and other essential molecules.

Transfer RNAs are small molecules which "translate" the language of mRNA into that of proteins. On one end of each tRNA is an anticodon loop, a short section of RNA that binds to a specific triplet of mRNA. At the other end is an amino acid. There are many types of tRNA within a cell, each associated with a particular amino acid and a particular codon of mRNA. When two tRNA molecules bind adjacent codons on mRNA, their amino acids are brought close together, ready to be joined as part of a protein.

The decoding of mRNA and the assembly of proteins is performed by ribosomes. These are themselves aggregates of rRNA (ribosomal RNA) and protein (not to be confused with the proteins that they help to produce). The ribosome acts as a "jig" to bring the right tRNAs together in the proper manner. The chances that one molecule of tRNA would spontaneously associate with its corresponding triplet codon on mRNA while floating about in the watery solution of a cell are slim. That two tRNAs would bind adjacent codons without the ribosome "jig" to guide them is almost inconceivable. The ribosome binds messenger RNA so that the mRNA codons are accessible.

Two tRNAs bearing amino acids are then attracted to two binding sites (the P site and the A site) on the ribosome. Their amino acids, now close together, both join to tRNA on the A site. The free tRNA (tRNA with no amino acid attached) leaves the P site. The ribosome then tracks three bases along the mRNA to the next codon, shifting the tRNA–amino-acid complex to the P site. In the empty A site, a new codon of mRNA is revealed. Another tRNA molecule comes to the A site and the process is repeated. The ribosome moves along the mRNA one codon at a time, and one amino acid at a time is added to the growing chain. The process ceases when the ribosome reaches a termination codon (either UAG, UAA or UGA). At this point, there is no corresponding tRNA to occupy the A site. The ribosome dissociates itself from the mRNA, ready to begin its work again. Many ribosomes can track along a single piece of mRNA. Each will translate the mRNA codes and produce an identical protein molecule. In bacteria, which have no nuclear membrane, transcription and translation take place virtually simultaneously.

A length of mRNA may code for one or several proteins. When it comes to the end of one protein-coding region, the ribosome may continue tracking along mRNA until it encounters another coding region. Protein synthesis will then recommence. The membranes of the endoplasmic reticulum which have channeled all the components of protein synthesis to the ribosomes also act as packaging for the completed proteins (and other molecules). The endoplasmic reticulum folds back on itself, enclosing the proteins in a membrane envelope, like a bubble formed from a soap film. Further processing of the proteins may occur in membrane vesicles as they transport the molecules elsewhere. An important property of transcription and translation is that they are not reversible. The flow of biochemical information is in one direction only.

The bacteria E. coli can produce a billion copies of itself in half a day

Gene cloning

It is one thing to know how something works, quite another to be able to alter it. Biochemists felt they had reached an impasse after the structure of DNA and the genetic code had been elucidated. They reasoned that if they could alter DNA sequences, they would be able to produce large amounts of rare proteins or new types of protein. These could be used for research or applications in medicine. In particular, they wanted to isolate and study genes and proteins from complex eukaryotic organisms. However, appropriate instruments for DNA surgery were not immediately available.

Almost all the research in the 1950s and 1960s on the function of DNA, RNA and proteins had been performed using bacteria like *E. coli*, or its viruses. These grew quickly and could mass-produce particular DNA sequences and proteins. For instance, *E. coli*'s single bacterial chromosome replicates each time the cell divides, a process which may take only 20 minutes. Each of its genes is duplicated. Eight copies can be obtained in an hour, 512 after three hours, and so on. After 10 hours, *E. coli* can produce 1,000 million exact copies, or clones, of each gene in its chromosome. Under the right conditions, any of these genes can direct the synthesis of proteins.

This biosynthetic power of microorganisms has now been harnessed through the development of gene-cloning technology. Fragments of DNA containing just a few genes can be taken from almost any source and placed in another organism. *E. coli* is very often used. The universal nature of the genetic code means that the bacterium treats the new DNA just like one of its own genes, producing thousands of millions of copies of the original piece of DNA. Yeast cells are also used as hosts for genetically engineered DNA. They do not grow as fast as *E. coli* but the proteins that they produce are often more like the natural product. For the same reason, animal cells are also used by genetic engineers. They are much more difficult and expensive to grow than microorganisms, largely because they are fragile and need special nutrients.

There are four basic steps in gene cloning. These are removal of a DNA fragment from the source using a restriction enzyme, insertion of the fragment into a vector (carrier molecule), introduction of the vector into *E. coli*, and replication of the cloning vector.

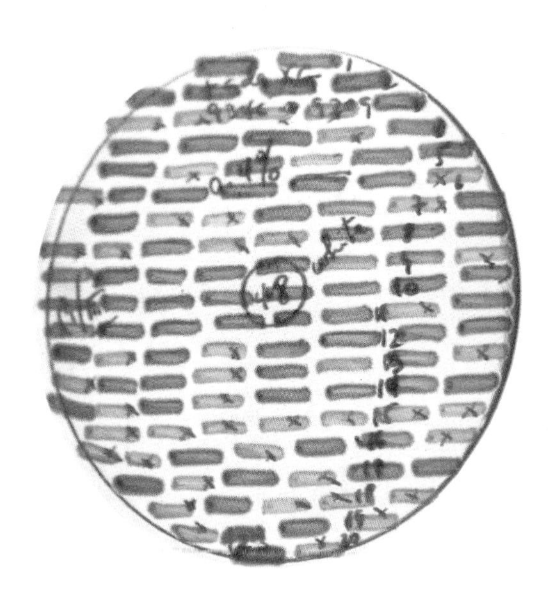

▼ Clones of genetically engineered E. coli. Each stripe, a colony containing hundreds of millions of identical bacteria, is derived from just one organism. Since DNA is invisible, genetic engineers have to use pieces of DNA which contain marker genes in order to keep track of what they are doing. One widely used marker gene is responsible for the blue color in some of the clones.

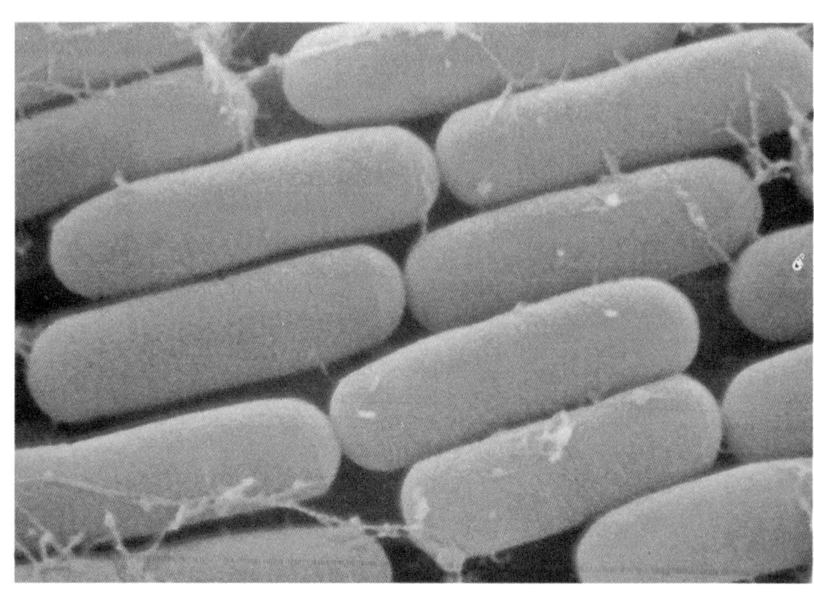

▼ Cells from a clone of E. coli. Although they vary somewhat in size, the millions of cells in a clone are all identical genetically. New DNA added to E. coli by genetic engineers, too, will be present in every cell of a clone. Genetic engineering enables scientists to use microorganisms like E. coli or yeast to produce an almost infinite number of replicas of any given piece of DNA, natural or artificial.

Chromosomes, genes and DNA

A chromosome is a large single piece of nucleic acid, generally DNA (some viruses have RNA chromosomes). Simple organisms like viruses have only one chromosome. More complex organisms have several. Normal human cells, for instance, have 23 pairs of chromosomes plus DNA contained in mitochondria (♦ page 7). In plants, there is also DNA from chloroplasts (♦ page 7). Even bacteria, such as E. coli, which have only one chromosome possess extra DNA called plasmids. In bacteria, chromosomes are circles of DNA, but in other organisms they are usually "linear" – that is, the two ends of a chromosome are not joined. DNA in chromosomes is usually tightly packed, wound tightly round itself in "supercoils". The separate chromosomes in a cell can be distinguished when stained with a special dye, but only at a particular stage in the cell's growth (♦ page 30).

The chromosomes contain genes. Genes are regions of nucleic acid which code for hereditary characteristics. This usually means that they direct the production of proteins. Some genes code for the sequence of the protein. Others regulate the transcription of DNA into RNA or the translation of mRNA into protein. Every gene codes for a function. There are other regions of nucleic acid (90 percent of the DNA in human cells) which appear to have no coding function. Non-coding DNA sequences occur not only between genes but also within genes. Those within genes are known as introns, and the coding regions that surround them are exons. Between genes are the so-called intragenic regions, some of which contain relics of disused genes (pseudogenes). There are pseudogenes, for instance, between genes which code for hemaglobin, These have sequences similar to the authentic genes but are truncated in form. The total gene content of an organism is its genome. The genome is the nucleic acid definition of an organism, a sort of genetic name.

DNA and RNA come in several forms. Short sequences are known as oligonucleotides (oligo : Greek, few). Hence chemical DNA synthesizers (♦ page 32), or gene machines, are also called oligonucleotide synthesizers. RNA is usually present in cells as single chains. However, certain viruses, including some that cause disease in rice and silkworms, have double-stranded RNA as their genetic material. Conversely, DNA is usually double-stranded or duplex, but certain viruses use single stranded molecules. In duplex molecules, the two strands are held together by hydrogen bonds (♦ page 15). The sequences of the two strands are complementary – an adenine in one binds to thymine in the other, guanine in one complements cytosine in the other. The strands can be separated by gentle heating to break the hydrogen bonds.

▼ Genetically engineered E. coli grown in sterile vessels at Biogen, one of the leading biotechnology companies. Experiments conducted in the glass reactor tank will enable the larger steel vessel to be used for growing large amounts of medically important proteins.

The discovery of restriction enzymes made gene cloning in the laboratory possible

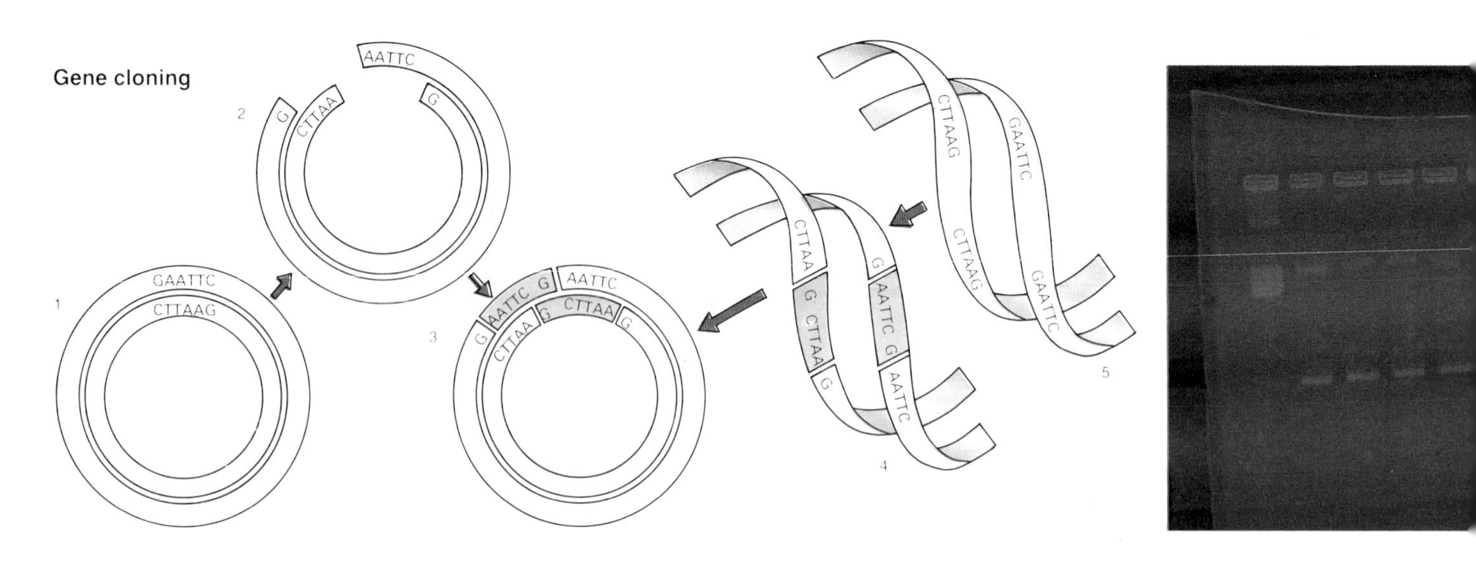

▲ *To clone a gene, genetic engineers add a small amount of a restriction enzyme to a test-tube of DNA containing the gene of interest (5). The enzyme cuts the DNA strand at two specific sequences to release the gene (4). A circular plasmid (1) is then cut (2) in one place by the same restriction enzyme and mixed with the DNA fragment. By virtue of the complementary "sticky ends" produced by the staggered cuts, the DNA fragment becomes inserted into the plasmid (3). The cuts in the DNA strands are closed by another enzyme, DNA ligase.*

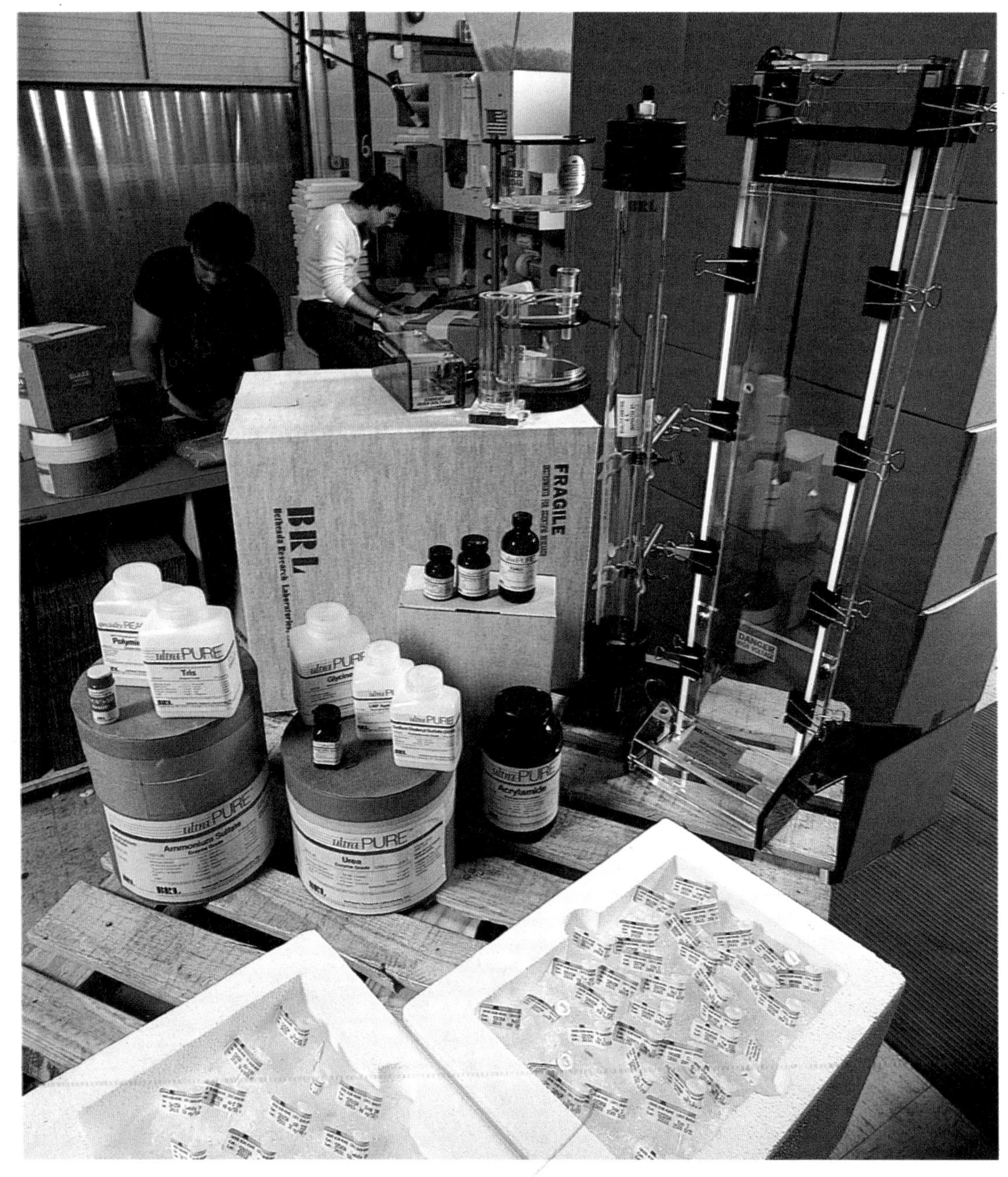

◄ *Ultraviolet light reveals the position of DNA bands in a gel, allowing them to be cut out one by one. Each band contains thousands of identical pieces of DNA, produced by treating an organism's chromosomes with a restriction enzyme. The DNA is later used in genetic engineering experiments.*

Cutting out the DNA

Until 1970 there were no convenient ways of fragmenting DNA. The molecule could be broken mechanically or biologically into very small pieces by one of several enzymes, but this could generate only randomly-sized DNA molecules. Such methods were not suitable for the purposes of cloning DNA. Then restriction enzymes were discovered. A particular restriction enzyme, when mixed in a test-tube with a given type of DNA, for example a human chromosome, would always cut the DNA at the same positions. Every time scientists repeated such an experiment, the same restriction enzyme gave the same result – a set of DNA fragments of different lengths but, each time, the same set. The restriction enzymes thus provided what was needed, an entirely predictable way of cutting up DNA. In scientific papers, genetic engineers refer to the enzymes by a form of shorthand that indicates their origin. For example *Eco*RI is an endonuclease obtained from *E. coli* strain R, while *Hae*II comes from *Haemophilus egypticus*.

◄ *Supplying restriction enzymes and other specialist reagents to genetic engineers is one of the most profitable parts of the biotechnology industry. A wide range of restriction enzymes is produced, each cutting at a different sequence. The enzymes are purified from cultures of microorganisms and often cost up to $100,000 per gram. Usually only very small amounts are needed, however, for cloning puposes. To ensure that the enzymes remain active, they are shipped packed in ice or "dry ice" (solid carbon dioxide) and very often delivered by courier.*

The sequence of the DNA determines where a restriction enzyme will cut. *Eco*RI, for instance, recognizes the six-base sequence GAATTC, *Hae*II, the four-base run, GGCC. Most important, a given restriction enzyme always cuts at precisely the same location, to the exact base, no matter what the source of the DNA. *Eco*RI breaks the phosphate bond in the DNA backbone between G and AATTC. *Hae*II does the same between GG and CC. Different enzymes may recognize different sequences of bases, or they may recognize the same sequence and cut at a different point within it or outside it. While the order of the bases dictates where a restriction enzyme will cut DNA, the length of the recognition sequence dictates how often cuts will be made. For example, an enzyme that recognized a single-base site would cut the DNA strand once every four bases on average. *Hae*II and other enzymes that recognize a four-base sequence, cut every 256 bases on average. *Eco*RI with its six-base recognition sequence cuts at every 2,000 bases or so.

Restriction enzyme sites have another important property. They are palindromic, in that the recognition sequence on one strand is the same as that on the complementary strand. This means that restriction enzymes cut both strands of duplex DNA. Where the cuts on opposite strands are staggered, the cleaved DNA will have short single-stranded ends which, in two pieces of DNA cut by the same restriction enzyme, will always be the same. These "sticky ends" are important in rejoining DNA after cleavage. In the same way that the two strands of DNA are held together by hydrogen bonds in the double helix, complementary sticky ends of DNA fragments will reassociate. The starting point of gene cloning is the digestion (cutting up) in a test-tube of DNA from the source organism, using one or more restriction enzymes. This creates a set of DNA fragments of varying lengths, only one of which contains the sequence of interest.

Sticky ends help to bring pieces of DNA together but they cannot hold them there permanently. An enzyme called DNA ligase is needed to join the fragments. DNA ligase can be purified from all living cells, where it plays a vital role in the replication of DNA. It closes gaps in the DNA backbone between single strands of DNA held in place by hydrogen bonding. It displays the same activity in the test-tube where it is used by genetic engineers to join the DNA molecules. Thus DNA can be both cut and rejoined in a controlled fashion. The next problem is to get the DNA into the host organism, *E. coli*.

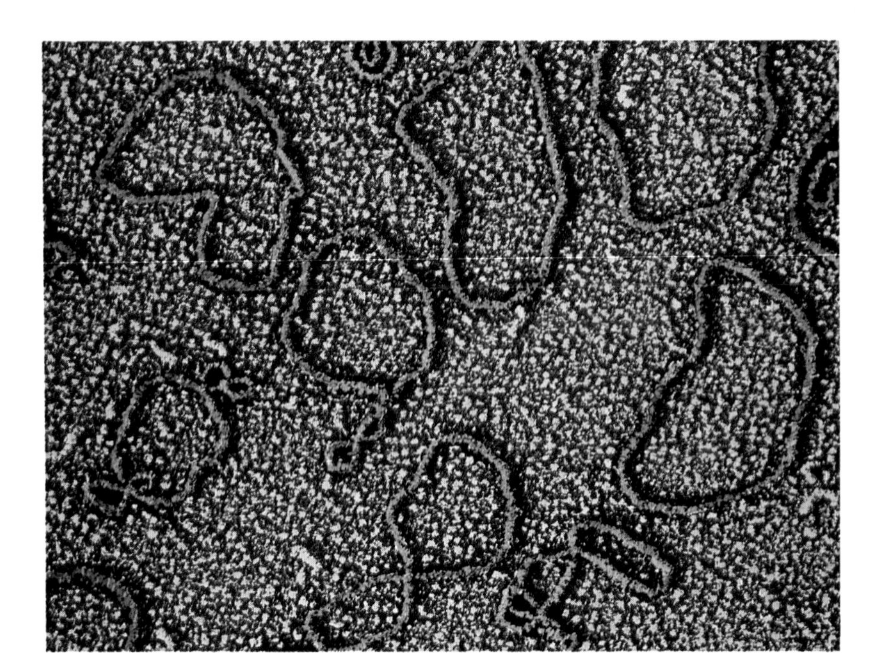

▲ *A view using an electron microscope of one of the first plasmids employed in gene cloning, pBR322, which comes from E. coli. Some of the plasmids are twisted around on themselves in what are known as supercoils. Plasmids are small circles of DNA found in many types of bacteria. They can pass from one organism to another or multiply within a cell. Often they confer useful functions, such as resistance to antibiotics, on their bacterial hosts. These properties make them suitable as vectors, carriers of DNA, for genetic engineers.*

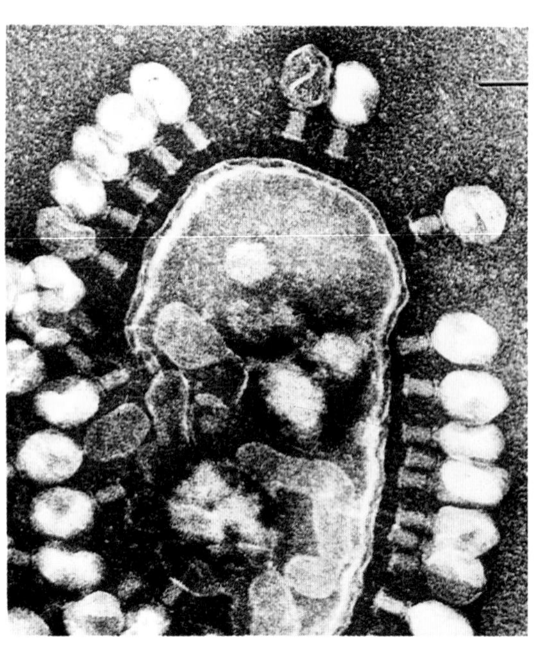

▲ *Scores of bacteriophages on the surface of an E. coli bacterium. The bacteriophages inject their DNA (contained in their bulbous "heads") through their tubular tails into the bacterium. Genetic engineers use modified 'phages as vectors, replacing nonessential 'phage genes with other pieces of DNA. The 'phage can carry as much DNA as will fit into its head. Empty 'phage heads can be loaded up with DNA in a test-tube with no bacteria present. 'Phage injection of foreign DNA into E. coli, is known as transfection.*

Vectors

Cloning vectors are the vehicles in which new fragments of DNA are transported into the host cells. The three main types include plasmids, bacteriophages (bacterial viruses) and cosmids.

Only small inserts of DNA (up to 10,000 bases) can be made before the use of plasmids becomes inefficient. However, this is more than enough for cloning single genes and their regulatory sequences (the average gene is about 1,000 bases long). More DNA (up to 24,000 bases) can be inserted in bacteriophage vectors. Many of these are derived from *lambda*, a simple virus which infects *E. coli*. To insert any more, essential functions would have to be deleted. This is exactly what genetic engineers do to produce another type of cloning vector, the cosmid. All the lambda genes are removed except the cos genes – those required for packaging DNA in the bacteriophage head. Thus very large DNA fragments (about 50,000 bases) have to be inserted before the vector DNA can be packaged.

The problem is to transfer the cloning vector constructed in the test-tube into an organism in which it can replicate. DNA borne by bacteriophage and cosmid vectors can enter the host cell by the efficient mechanism of "phage infection". The transfer of plasmid DNA is much less efficient. Plasmids are not conveniently packaged in protein coats. Circles of plasmid DNA have to enter *E. coli* by a mechanism known as transformation. In transformation, plasmid DNA simply adheres to the cell surface and penetrates the cell. Many other organisms including bacteria, mammalian cells, and plant cells can be transformed by DNA only after chemical, enzymatic and electrical pretreatments. Transformation is extremely important in gentic engineering because it is a nonspecific method of transferring DNA. DNA from almost any source can be introduced into suitable cells.

Methods very similar in principle have also been developed for cloning genes in other bacteria, yeast, fungi, plants and animals. Genetic engineers have thus effectively broken down the barriers between species. If suitable DNA cloning vectors are available, the genetic material from any organism can be introduced into any other. Indeed, artificially created sequences of DNA that have never existed in nature can also be transferred.

The Information of Life

2

The language of DNA...Reading and writing DNA...DNA sequencing...Gene machines...DNA databanks...Gene libraries...DNA and evolution...Genetic conservation...DNA as messenger...Defective genes...Molecular fingerprinting...The cell as information processor

▲ *Working with X-ray pictures of gels, a scientist compares the relative positions of bands of DNA in sequencing "ladders". This enables him to establish the order of bases in a piece of DNA. Japanese scientists have developed a highly automated method of DNA sequencing in which the gels are very thin layers on top of photographic film. The film can then be scanned mechanically.*

▼ *DNA sequences are compared with the aid of a simple computer program. The transparent mixing chamber of a "gene machine" used to determine sequences rests on a homology chart. The sequence of one gene is plotted down the page, that of another across it. Every time the sequences coincide, the computer draws a short line. The lines are joined up where regions are similar.*

Reading DNA

Scientists represent the bases of DNA as letters – A, T, C and G. Genes then become long words and can be analyzed efficiently by computers. To read the language of DNA it is necessary to determine the order of bases in the chain. There is no direct way of doing this. Sequencing, therefore, relies on converting the order of bases into something that can be measured. Scientists have an accurate way of comparing the lengths of DNA fragments. It is a process called electrophoresis, which separates different-sized molecules electrically. DNA is a negatively-charged molecule (due to the phosphate groups in its backbone) and in an electric field will migrate towards the positive electrode. Smaller fragments of DNA always travel more quickly than larger ones. If a radioactive compound is attached to the molecules, they can be detected using X-ray film. By comparing the distances that the fragments migrate, scientists can distinguish molecules of DNA with one, two, three and four bases, and so on up to 500 bases.

But how is electrophoresis adapted to yield information on the sequence of DNA molecules rather than just their lengths? The trick is to produce DNA fragments in a sequence-specific way. First, a radioactive phosphate molecule is inserted at one end of a single-stranded DNA sequence (after the two strands of the double helix have been separated by heating). Next, four chemicals are used to disrupt the DNA chain at each of the four bases. One chemical breaks the DNA chains before guanine bases, another cleaves them before cytosine, and so on. Each of the four chemical treatments produces a different set of fragments. Electrophoresis can then be used to compare fragment lengths, and from the pattern of radioactive bands scientists can work out the sequence of the original DNA. Thus, if a molecule six bases in length is produced by the chemical which cleaves before guanine bases, then the seventh base must be guanine. The same chemical might produce fragments of 10, 11 and 16 bases, indicating guanines at position 11, 12 and 17. In a similar way, patterns are built up for the other bases. When all four patterns are combined, the DNA sequence is revealed.

Sequence-specific DNA fragments can also be produced biochemically using compounds called dideoxynucleotides. These are identical to the normal components of DNA, the deoxynucleotides, except that they are missing an oxygen atom. They can "fool" enzymes into incorporating them into growing strands of DNA. However, once in the strand, the absence of the oxygen atom prevents the DNA from growing any further. By blocking DNA synthesis with different dideoxynucleotides, one corresponding to each of the four DNA bases, sequence-specific fragments of different lengths are obtained. These can be analyzed by electrophoresis.

These DNA sequencing methods were developed in 1977 by Walter Gilbert (b.1932) and Allan Maxam at Harvard University and Fred Sanger (b.1918) at Cambridge, UK. Both methods require only small amounts of purified DNA. Five millionths of a gram (5 micrograms) is sufficient for one sequencing experiment and can readily be produced by DNA cloning methods (◀ page 24–28). Scientists have sequenced thousands of different genes and many longer DNA chains such as those of complete viruses and plasmids. Researchers now talk of being able to "read" DNA. The development of automated sequencing machines (sequenators) has meant that DNA containing hundreds of bases can now be sequenced every week.

The most ambitious plan so far is to sequence all the DNA in the human cell, the entire human genome. Each cell in our body contains 23 pairs of chromosomes, together coding for every aspect of our biological existence. In these chromosomes there are approximately 3,000 million base pairs of DNA. At a rate of about 200 bases per week, it would take 600 scientists 500 years to obtain the complete sequence. To tackle the human genome, advanced automatons able to sequence 15,000 bases of DNA per week are being developed. With 500 such machines, the human genome could be sequenced in eight years. Different teams of scientists will sequence simultaneously, starting at various points on each of the human chromosomes.

To ensure that they are not duplicating effort, scientists will need to refer to a "map" of human DNA. Like all maps, this will represent the relative positions of easily-observed features, which in this case will be physical or genetic. The chromosomes of cells, seen under a microscope and treated with appropriate drugs, form worm-like shapes which scientists can recognize and map. Genetic maps of DNA are based on patterns of inheritance and indicate the order of genes or chromosomes. Genes that are closer together are more likely to be inherited together than those that are farther apart. By comparing inheritance rates, scientists can work out how close genes are to each other. Using genes and other DNA features (such as restriction enzyme cutting sites) as reference points on a DNA map, scientists can relate one DNA sequence to another. Maps enable DNA sequencing efforts to be sensibly coordinated.

▶ Special stains are used to reveal distinctive patterns of dark and light banding in human chromosomes. Geneticists use these patterns to distinguish the 23 different pairs of chromosomes of human cells. However, they can also be used to map the positions on the chromosomes of particular genes. For instance, patients with leukemia may have a gap in chromosome 22. By correlating such observable changes in patients' chromosomes with genetic changes, scientists can roughly locate the positions of certain human genes.

▶ The first step toward the sequencing of the entire length of human DNA is to construct a rough "gene map" which gives reference points for further studies. Here, the four grandparents, both parents and nine children of one extended family are told by a researcher how tracking the inheritance of their genes will help in mapping DNA. This research is being conducted in Salt Lake City, Utah, among the Mormons who keep very detailed family records.

▶ A combination of electron microscopy and genetics enable scientists to produce physical maps of genes. The yellow circle of double-stranded DNA has been mixed with single-stranded RNA. The RNA has annealed to one strand of the DNA (red), creating a loop which reveals the position of the gene being mapped (blue).

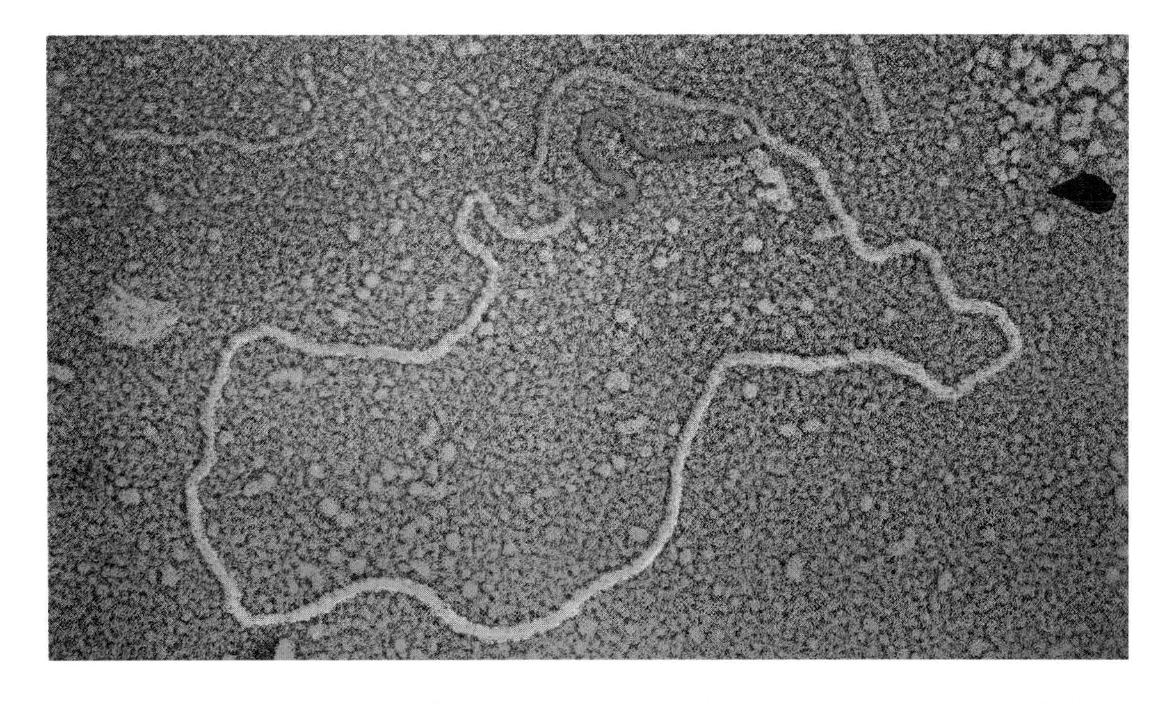

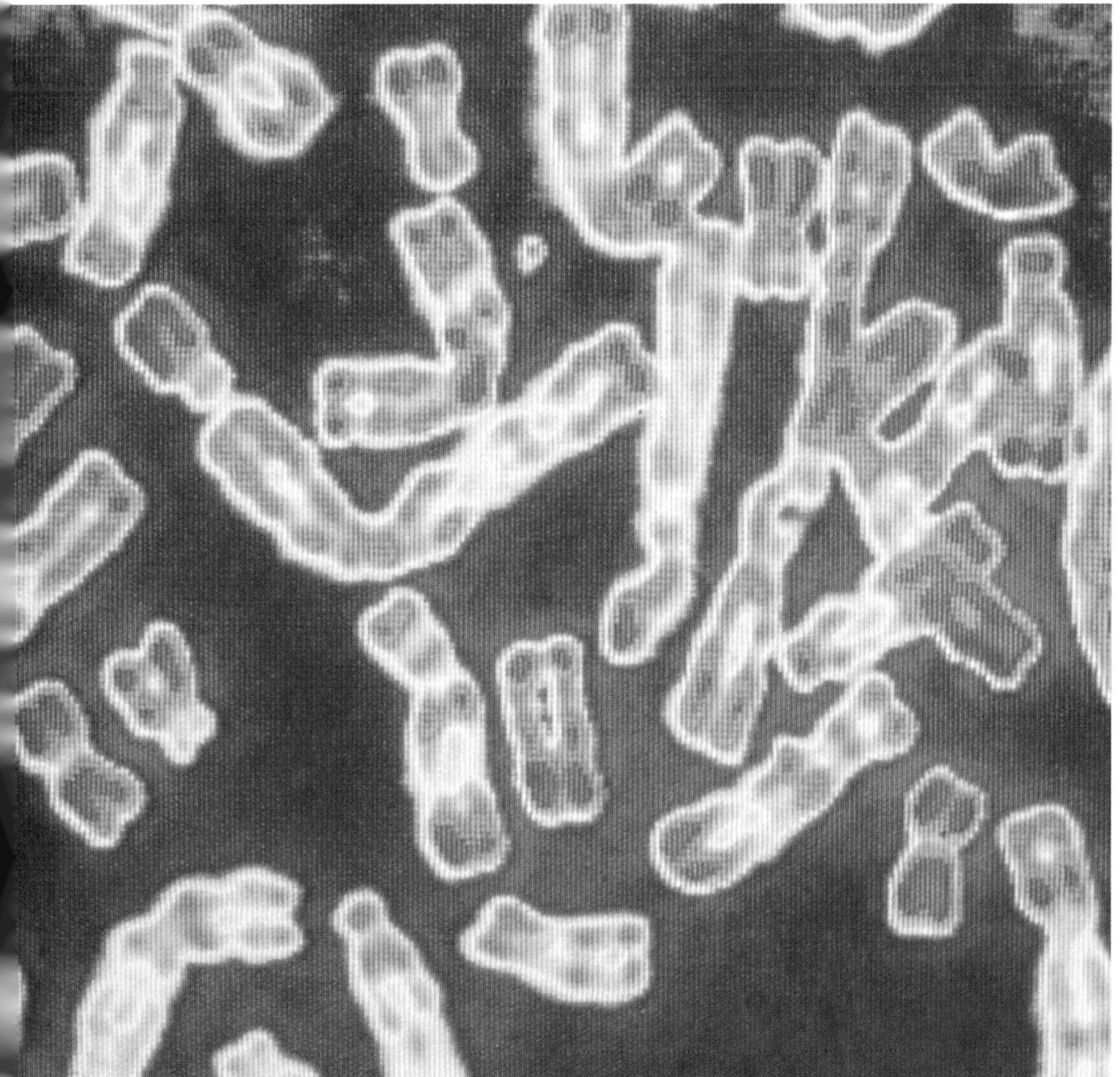

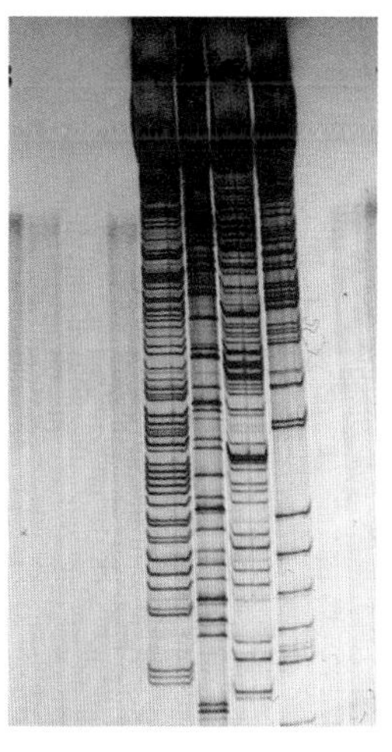

▲ ***Sequencing the human genome is a huge project. X-rays of DNA-containing gels, such as this one, allow scientists to determine the sequence of pieces of DNA containing 300 or so bases. At least 10 million such gels would be needed to span human DNA. The first task for scientists, therefore, is to develop rapid sequencing methods that can be automated.***

DNA sequences are built up base by base, in the same way as children spell out unfamiliar words by stringing together the sounds represented by the individual letters. Once a word has been learned, however, a child will recognize it on a future occasion. DNA's equivalent of word recognition is probe technology. If two strands of DNA, held together by weak bonds (hydrogen bonds), are separated by gentle heating to give single-stranded molecules, the hydrogen bonds will reform, and the complementary strands pair up, when the DNA cools again.

This phenomenon was first observed with large pieces of DNA. Researchers now use as probes short sequences of DNA containing as few as 20 bases to search for particular regions of long DNA sequences. The short probes can be synthesized chemically, or longer ones can be obtained by gene cloning. Probes "hunt" for a complementary sequence in a DNA sample (like the word-finding function of a word processor, or a mainframe computer looking for a particular DNA sequence) and bind to it. The difference is that it is real molecules of DNA that are being scanned, not just the letters that stand for them. Clinical samples of blood, for instance, can provide the DNA for the test.

The recognition and binding of part of a DNA molecule by a probe takes place silently and invisibly. Biotechnologists have to make this interaction detectable. One method is to use radioactive DNA probes, produced by growing the organism containing the probe DNA on radioactive biochemicals.

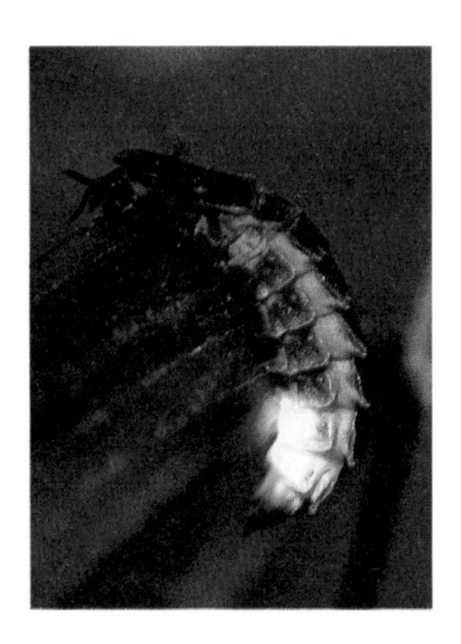

◀ Enzymes in glowworms (far left) and fireflies convert chemical energy into light to create spectacular natural displays. Biotechnologists use light-producing enzymes called luciferases from fireflies to replace radioactive tags in DNA assays. They attach luciferase to short pieces of DNA (probes) and then mix the probes with sample DNA. If the probe binds to a complementary sequence, luciferase produces light when chemicals are added. The light is detected photographically or with a luminometer.

▲ *Gene machines are just automated chemistry sets. The machine pumps a precise amount of one of four solutions of the bases of DNA contained in the jars along fine pipes to a reaction chamber. A computer controls which base is added. The base is added chemically to a growing chain of DNA. The addition/reaction cycle repeats until the entire sequence is made. In this way, scientists can alter DNA base by base and study the changes in proteins and organisms.*

After sample DNA is extracted from its source, it is converted to a single-stranded form, usually by a combination of heat and chemical treatments. It is then transferred to a nitrocellulose filter, a sort of brittle blotting paper. The nitrocellulose filter is subsequently soaked in a solution of the probe for several hours so that binding can take place. The filter is then washed, leaving any probe present attached only to the DNA sample. The presence of the probe is detected with X-ray film, which is fogged by radioactivity. Radioactive DNA probes are being replaced with probes labeled with fluorescent chemicals.

Probes can detect very tiny amounts of DNA, down to one tenth of one millionth of one millionth of a gram (0.1 picogram). Consequently probes have a large number of medical uses (➧ page 91). DNA probes are also the basis of a technique known as molecular (or genetic) fingerprinting, used in police work (➧ page 42).

Writing DNA

As well as being "read" by sequencing or DNA probes, DNA can also be "written", using oligonucleotide synthesizers. Sometimes called "gene machines", these are computer-controlled instruments that can be programmed to manufacture short lengths of single-stranded DNA. Gene machines, however, enable scientists to produce *any* DNA sequence, not just those found in nature. DNA produced in gene machines can be inserted into living organisms using genetic engineering techniques. The artificial genes will be processed by cells in exactly the same way as their own DNA.

Even the largest computerized databank will contain only a fraction of the genetic information of life on Earth

Storing DNA

DNA sequence data stored in a computer-accessible form, as long strings of the four letters A, C, G and T, are invaluable to scientists. Powerful mainframes can search through an entire database looking for a particular sequence of DNA or its close relatives in the same way that a home computer operating as a word processor can find specified words in a long piece of text. Sequence comparisons have enabled scientists to understand how genes are constructed and controlled, and have led to discoveries about the causes of inherited diseases and their prevention. An important clue in understanding cancer was unearthed when a sequence of DNA often found in cancer cells was discovered to be almost identical to that of a virus. To cope with the burgeoning volume of sequencing information, DNA sequence libraries, called GenBank, were set up in Los Alamos and Heidelberg, West Germany.

Biological information can also be stored in computers as strings of one-letter amino acid codes representing proteins. Methods for determining the order of amino acids were developed more than a decade earlier than DNA sequencing. Much of our knowledge of evolution has come from comparisons of protein rather than DNA sequences. The sequences of around 6,000 different proteins corresponding to about one and a half million amino acids are known. However, this is much less information than is available from DNA sequences, partly because only certain regions of DNA give rise to proteins. In addition, automation has made DNA sequencing much faster than protein sequencing. It is quicker to clone and sequence a gene and use the genetic code to convert this into an amino acid sequence than it is to sequence a purified protein.

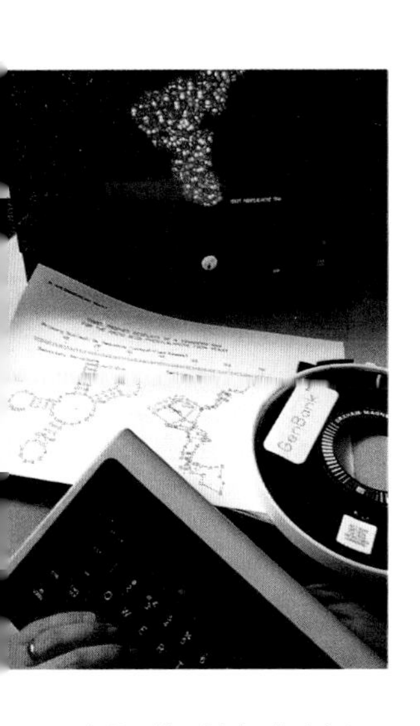

▲ *Banking biological data-withdrawals. For a small fee, scientists can buy DNA sequence information on magnetic tape from the computerized library GenBank. In 1986, the databank was storing 11 million bases, but by 1995 this may rise to a billion, with millions more being added each week.*

◄ *Banking biological data-deposits. A researcher analyzes the results of a DNA sequencing experiment and enters them directly into a desk-top computer. When the results are published in a scientific journal, the scientist may submit the sequence to GenBank on floppy disks, magnetic tape or by electronic mail, so that fellow researchers can have ready access to it.*

▲ *A display of a specialized database, established by the US biotechnology company Cetus, to help researchers find the few scientific papers that interest them among the tens of thousands published each year. The entire database can be scanned in seconds by entering a few keywords.*

Nevertheless, even the largest computerized databanks will contain only a fraction of the genetic information which exists on Earth. Three billion bases – the information that may be gathered in the next 20 years – equates to less than the genetical material of two species of mammals or insects. There are about 4,500 mammalian species known and at least one million insects. It would be difficult, though not impossible, to conceive of ways in which the information of the biosphere could be stored. But the real bottleneck is in converting genetic data stored as DNA into a nonbiological form. To avoid this, scientists are using simple organisms to store biological information.

A DNA library is a collection of DNA fragments which, ideally, represents all the sequences of a single organism. To produce a DNA library, an organism's DNA is purified and then cut into manageable fragments using restriction enzymes or mechanical shearing. Each of the fragments is inserted into a piece of carrier DNA, or vector, so that it can be cloned in a bacterium (usually *E. coli.*; ◄ page 24). Cloning allows scientists to obtain billions of exact copies from a single piece of DNA. This gives them enough material to work with.

The size of a library depends on how much DNA there was in the original organism and on the size of the average insert in the vector. Thus a mammalian genome with approximately three billion base-pairs of DNA would require at least 500,000 plasmid clones or 60,000 cosmids (◄ page 28). The smaller chromosomes of bacteria require 1,000 times fewer vectors. In practice, a complete genomic library will have several times as many clones as this to ensure that vital regions of DNA have not been dissected during preparation of the clones. In cDNA libraries (c stands for copy) DNA copies of messenger RNA are made; cDNA libraries are smaller than genomic libraries and contain only DNA for genes.

A DNA library's primary purpose is to allow genetic engineers to retrieve particular pieces of DNA, though such libraries can also be used to store genetic information. Unlike a conventional library, a DNA library is not organized. To find a particular DNA sequence the contents of the clones have to be examined. Up to 100,000 separate clones can be preserved in a deep freeze on a circle of filter paper the size of a standard floppy disk. One filter disk contains up to 1,000 megabytes of DNA information, a thousand times more than the densest floppy. The entire human genome could be contained on one piece of paper.

The DNA of an organism is present within the organism itself in an even more economical form. We are all repositories for our own chromosomes. For some species, this is a convenient method of DNA storage. Many strains of bacteria, yeast and fungi, for instance, are maintained in culture collections. The precise method of storage depends on how much is known about the organism, its growth requirements and survival mechanisms. Microorganisms can be deep-frozen in sugary solutions, dried and kept at room temperature, or preserved in liquid nitrogen. If none of these treatments is appropriate, the organism can be cultivated over an indefinite period in sterilized conditions. Microbial culture collections worldwide contain hundreds of thousands of organisms. These have assumed a legal importance. Patents for microbial inventions require the organism to be deposited in several international culture collections.

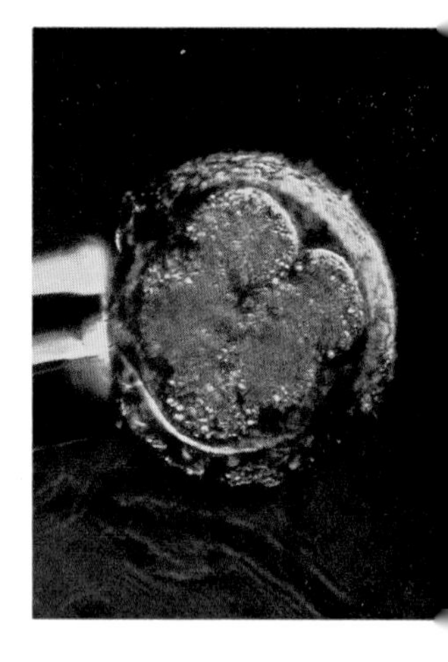

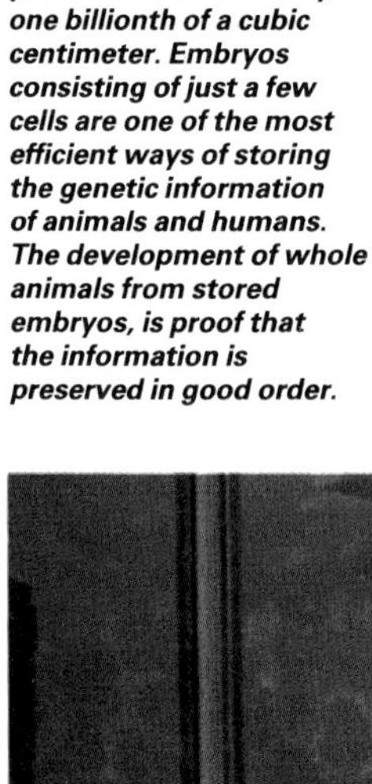

▶ Each cell of the bovine embryo contains around 3 billion bases of DNA, and yet has a volume of only one billionth of a cubic centimeter. Embryos consisting of just a few cells are one of the most efficient ways of storing the genetic information of animals and humans. The development of whole animals from stored embryos, is proof that the information is preserved in good order.

Single cells of many eukaryote (◀ page 6) species can be grown in much the same way as microorganisms. Freezing and liquid nitrogen are used as storage media for the cells of many plants and animals including humans. Sperm banks, for instance, have been an integral

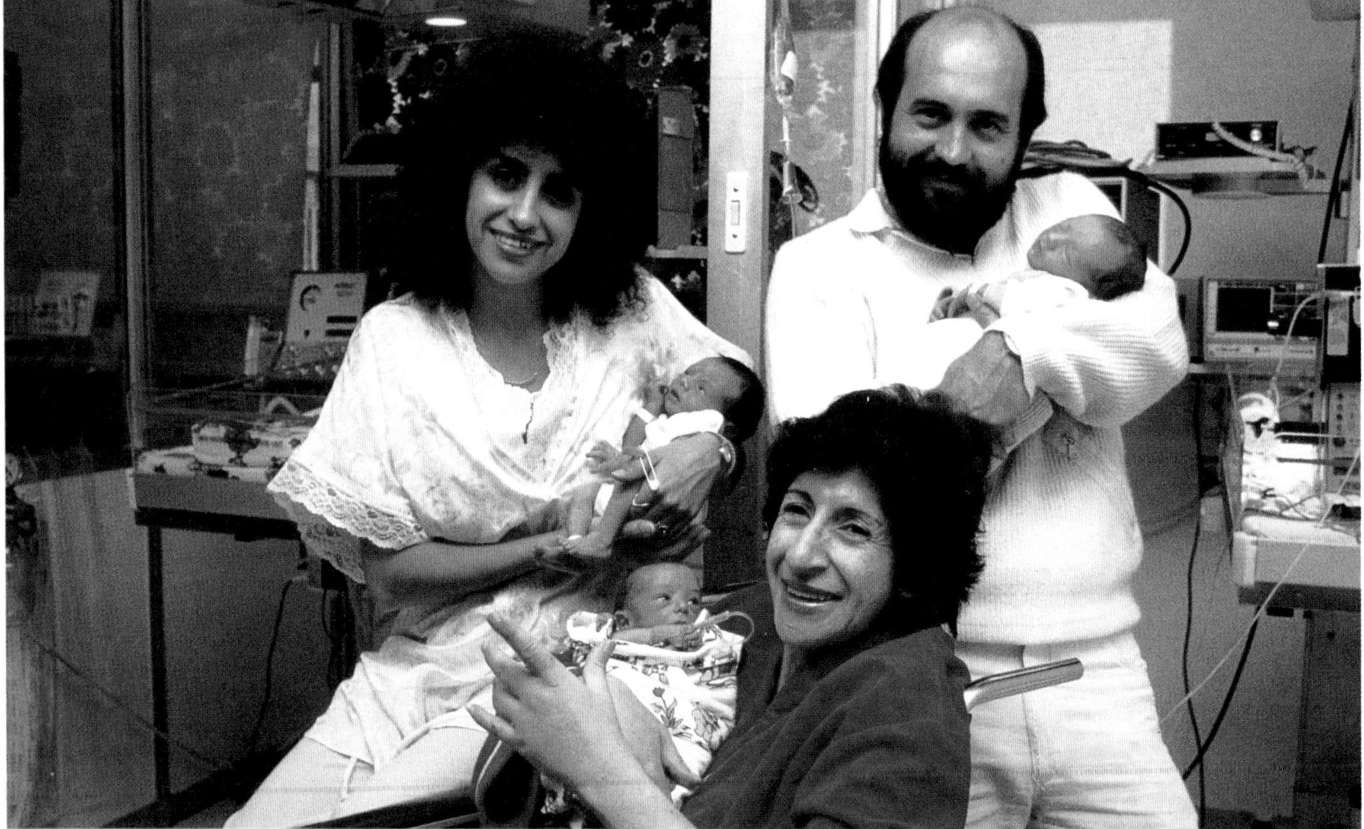

▲ *Embryos being stored at –190°C in liquid nitrogen. At this low temperature, embryos can be preserved in perfect condition almost indefinitely. In the UK, a pair of twins were born over two years apart when, following in vitro fertilization, one embryo was implanted immediately and a second frozen for later implantation.*

◀ *Pat Anthony (seated), a 48-year-old South African gave birth to her own grandchildren in 1987. Eggs from her daughter Karen were fertilized by sperm from Karen's husband, using test-tube techniques. Four embryos were implanted in Pat. Three of them developed into David, Jose and Paula. Pat is legally their mother, genetically their grandmother.*

▲ *For many rare species like this gorilla, preservation of genetic material in liquid nitrogen in an "ark of the 21st century" may be the only way to save them from extinction. Such measures may provide time for humankind to ponder the implications of its incursions into unique habitats.*

part of artificial insemination programs in agriculture since the 1950s. They have allowed wide geographical distribution of elite genetic material in livestock and in more exotic animals such as the gorilla and giant panda, which are endangered species. Artificial insemination is often available to human couples as treatment for infertility. Moreover, human sperm from an "elite stock" of Nobel prize winners, Olympic athletes and prominent figures is available commercially for those who feel their own genetic material is inadequate.

Also important is the storage of egg cells and embryos. Techniques such as in vitro fertilization use genetic material preserved in this way. The ability to control reproduction by storing embryos under laboratory conditions is a boon to the farmer and the zoologist. The female line of elite animals can be used to improve livestock. Storage of human embryos, however, creates a reproductive time-warp. Twins conceived at the same time have been born years apart. A South African woman gave birth to her own grandchildren after the triplets conceived in vitro by her daughter and the daughter's husband were implanted in her womb. For several years, a freezer in Australia contained embryos which were heirs to a huge fortune left by their parents who were killed in an air crash. The Australian government has now decreed that the embryos should be donated to medical research.

Artificial storage of DNA is perhaps the last hope for the conservation of many species on the brink of extinction. San Diego zoo in California has a large-scale liquid nitrogen tank in which are preserved the DNA of very rare species. This "ark of the 20th century" may be the only source of genetic material of some organisms, as the 21st millenium approaches.

DNA and evolution

Why are horses and zebras clearly closely related while moths and mosses obviously have little in common? The basis of these similarities and differences is the variation in DNA sequences between organisms. Exploration of existing DNA sequences reveals an enormous amount of information about the way life has evolved from the earliest DNA-containing organisms.

The rate at which DNA changes – its mutation rate – is rather like a clock. In an average gene, the DNA timepiece only ticks once every million years or so, when an acceptable mutation (one that does not cause the death of the organism) arises. Thus the present form of a gene which has been evolving for 20 million years would be expected to differ from its original version in about 20 places. By looking at gene sequences from two species, scientists can estimate how long ago the last common ancestors of the two species existed. They can then construct family trees to explore the relationships between species. Fossil evidence suggested that the last common ancestor of humans and the African apes lived 20–30 million years ago. However, by comparing the structures and sequences of blood proteins from the two species, molecular biologists in California demonstrated that humans and apes had a common ancestor as recently as five million years ago.

Using very slow running DNA clocks, it is possible to go much further back and trace the history of bacteria. Sequence comparisons of ribosomal RNAs (◆ page 22), molecules which have evolved very slowly, show that closely related bacteria from the human gut, like E. coli and Salmonella, diverged about 140 million years ago. Going back a further 1,000 million years or so, a common ancestor of bacteria and mitochondria can be traced. The emergence of the mitochondria, the powerhouses of eukaryotes, represents the genesis of all higher organisms. The chloroplast, the light-trapping organelle of all plant cells, is even more ancient. It and all true bacteria developed from a common primitive form which existed about 1,500 million years ago. With DNA as a molecular clock, the development of life can be traced back over a period of 3,000 million years. On the basis of DNA evidence, scientists believe that life arose only once on this planet.

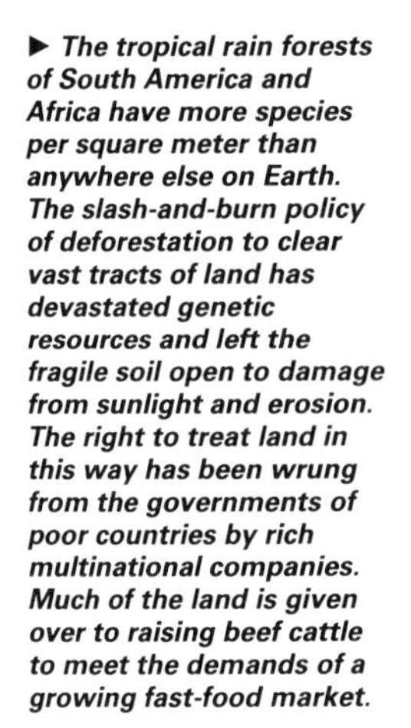

▶ The tropical rain forests of South America and Africa have more species per square meter than anywhere else on Earth. The slash-and-burn policy of deforestation to clear vast tracts of land has devastated genetic resources and left the fragile soil open to damage from sunlight and erosion. The right to treat land in this way has been wrung from the governments of poor countries by rich multinational companies. Much of the land is given over to raising beef cattle to meet the demands of a growing fast-food market.

Organisms and organization

As human beings explore further into the hidden reaches of the Earth, an ever-increasing variety of living creatures is uncovered. Life comes in an immense array of colors, sizes and forms and is found in almost every imaginable environment – from steamy, high-pressure sulfurous vents under the oceans, where temperatures exceed that of boiling water, to the icy wastes of Antarctica. Yet, despite its diversity, the biological world is far from chaotic. The attributes of each creature are ordered by the demands of its habitat.

For four thousand million years organisms have been diversifying. During this time, many thousands have disappeared, unable to find an appropriate niche. In the 20th century, human expansion has become, for the first time, the major cause of extinction. Agriculture consumes more and more land for the cultivation of only a few varieties of plant. Three crops – rice, wheat and maize – account for half of the world's arable acreage. Only 200 of 240,000 known plants feature in agriculture. The same picture is seen with animal species. Huge areas of the Brazilian rainforest are being cleared by slashing and burning as pasture for beef cattle. Formerly, the forest supported hundreds of different species per acre. Now the land perpetuates just one.

▲ *The beauty of many tropical plants, such as the flowering black bean tree, is only part of their attraction. Every species has unique attributes and although these have evolved to meet the needs of the plant, some of them may be useful to humans. Many wild relatives of agricultural crops, for instance, are more resistant to disease or to drought than cultivated varieties. Plants produce the widest range of complex chemicals of any group of organisms. Many of these might be used as medicines, flavorings or colorings.*

Human beings are not conspicuous in their attempts to halt or reverse the contraction of living diversity. Instead they are banking biology. During the 1980s many developed countries sent genetic missions to tropical regions, which have the greatest abundance of species, to collect DNA. Plant species, on which people have depended for nutrition and medicine for thousands of years, are a particular target for the teams of scientists from Japan, the USA and Europe. These activities are an investment for the future. As organisms become rarer, like paintings their value will increase. Some of the species will be used by biotechnologists to yield valuable products (for instance, a plant called *Lithospermum erythorhizon* which had almost become extinct, is now grown as a plant cell culture in Japan to produce shikonin, a dye used in cosmetics and pharmaceuticals). Yet for every one collected, tens if not hundreds will be lost for ever as human beings forge relentlessly ahead with their own future.

The Earth is the richest store of biological information, inhabited by between 5 and 30 million species. This is 4.9–29.9 million more species than are stored currently in zoos and culture collections. On average, organisms contain around one billion bases of DNA. If all members of a species had the same DNA, then the entire biosphere would provide 5–30 million billion bases of DNA in different arrangements. (Actually, there are millions of times more arrangements than this – all members of the human species, for instance, have slightly different DNA.) However, not all DNA is useful. Only 10 percent of DNA in mammals codes for protein. Therefore, for storage purposes, perhaps only a half to three million billion of the Earth's DNA bases need be considered. Furthermore, many of the useful sequences are present in many species. In extremely closely related species like humans and chimpanzees, only 1 percent of the DNA is different. If this figure is applied generally, the Earth's entire genetic diversity might be represented by as few as 5,000–30,000 billion DNA bases. Yet, this information would require 1,000 computerized DNA databanks each 500 times bigger than those existing currently. However, these statistics perhaps obscure the main point. The preservation of the Earth's genetic diversity needs organisms, not human organization.

The genetic messenger

DNA is a message which passes through time from one generation to another and contributes to the development and appearance of the individual. Family groups often share features: a prominent nose, red hair, skin color, height. These common attributes reflect a common genetic makeup among blood relatives. Identical twins have identical genes. Other brother and sister pairs share half their genes. We inherit half our genes from each parent. We have a quarter of our DNA in common with each grandparent and with each blood uncle or aunt.

In our complex societies, it sometimes becomes necessary for people to be able to prove that they are related. Increasingly, the double helix is being called to the witness stand in legal proceedings to provide this proof. In Britain and in Canada, stricter immigration laws have led the authorities to use genetic means to check whether newcomers to the countries are closely related to existing residents. DNA extracted from blood samples of, for instance, parents and children can be compared by genetic fingerprinting (⧫ page 42). The analysis will give almost conclusive proof of any parent–child relationship, certainly more convincing than complex blood tests. Similar tests have been used in paternity suits to establish parenthood.

Genetic analysis can also identify lost children when families have been separated by natural disasters, war or other violence. During the recent period of military rule in Argentina, thousands of citizens were abducted, never to return. Many of "the disappeared" were parents, including some pregnant mothers. They are now dead, but the fate of the children is less clear. An Argentinian human rights group called the Grandmothers of Plaza de Mayo believes that many of the children

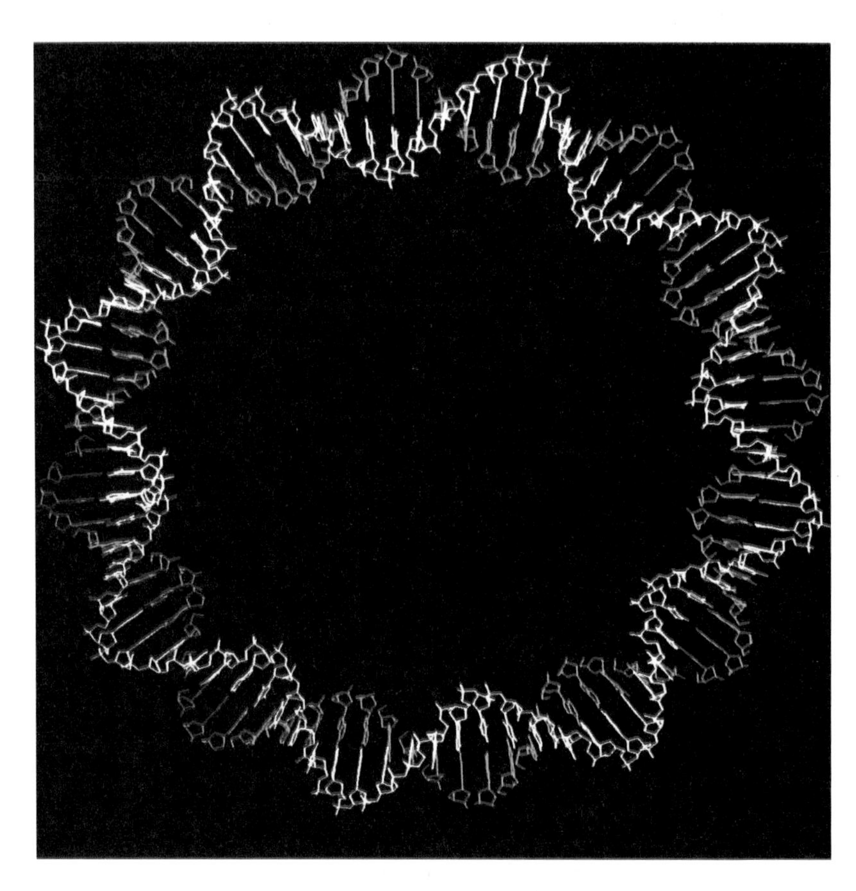

▶ DNA has many stories to tell. For instance, it reveals breeding. Whether humans or horses, some of the DNA of the offspring is identical to that from the parents. With DNA probes, scientists can resolve paternity suits or test whether foals are thoroughbreds. Even inanimate materials can have an invisible DNA identity. The FBI is said to be impregnating all its photocopy paper with unique DNA sequences to trace the source of any leaked documents.

are still alive, "adopted" by military personnel who may have been involved in the murder of the child's natural parents.

With the parents dead, establishing relationships depends on comparing genes from the child with those of the grandparents. If two or three of the grandparents are still alive, this can be done by an advanced form of blood typing. But if only one grandparent remains, analysis of DNA is required. Since a child inherits a quarter of its DNA from each grandparent, DNA probes are used to look for inherited sequences in the child's genes. Scientists in Argentina are building up a genetic bank of those grandparents whose grandchildren are missing, which will permit identification even after the grandparents are dead. As a result of genetic analysis, some 40 children of "the disappeared" have been reunited with their true families.

Our own DNA may contain messages that we would rather not receive. Oncogenes are short regions of DNA in our normal cells which can be triggered to convert the cell to cancerous growth. The triggers – carcinogens – alter the gene or its regulating mechanisms and uncontrolled cell growth occurs. Carcinogens include chemical agents such as cigarette smoke, radiation from, for instance, the Sun, and certain viruses. Oncogenes are not defects in our genetic makeup. Normally, they code for essential metabolites like growth factors.

◀ ***The suppression of information, which often accompanies the suppression of human rights, cannot silence the messages contained in DNA. When civil law replaced the military regime in Argentina, many people sought their relatives who had gone missing. Photographs of the "disappeared" were shown on television and paraded on marches. If young children had been abducted and then adopted, it was possible to find them but difficult to prove who they were. When scientists conducted genetic tests, however, there could no longer be any argument.***

Other unwelcome DNA messages in our genes are defects. Diseases such as hemophilia, cystic fibrosis, Duchenne muscular dystrophy, and Huntington's disease are caused by single coding errors in our genetic material. Down's syndrome sufferers have three copies of chromosome 21 instead of two. Even for conditions such as thrombosis and rheumatoid arthritis, genetic defects contribute to the disease although diet and proper healthcare are also important.

We frequently receive unsolicited DNA transmissions from other organisms. These are the infectious diseases. Viruses are particularly intrusive. They release their DNA or RNA into our cells to transform our cellular workings into machinery for making virus nucleic acids and proteins. Once that task is complete, the virus invades a fresh cell and starts again. The body has developed several mechanisms to repel or contain such genetic invaders but these are not always effective. Millions of people die each year from infectious disease. Infectious messages spread in epidemics. AIDS (acquired immune deficiency syndrome), is caused by the human immunodeficiency virus. When the viral RNA enters human immune cells, it converts to DNA, which is then integrated into host chromosomes. In Africa the disease now affects an estimated one million men, women and children.

▶ ***The AIDS virus spreads its genetic message through the immune cells of infected people and from person to person. However, soon after AIDS was recognized as a problem, scientists decoded much of the sequence of the virus's RNA, allowing earlier diagnosis of the disease to reduce the rate of its spread. At the same time, genetic sequences of the AIDS virus were put into other microorganisms enabling the production of viral proteins as potential vaccines.***

Using the same principle, genetic engineers can endow their products with DNA trademarks. Many of the microorganisms produced by biotechnology are difficult to develop. Intricate genetic manipulation and product improvement require much research time and expensive equipment. Yet living products, especially those used in the open agricultural environment, are easy to copy. An unscrupulous firm could simply take a spadeful of soil, or a plant or an animal, from a field and obtain the biological products of its competitors. To protect their investment, many companies will incorporate genetic trademarks into their organisms. The trademarks could be random sequences of DNA bases inserted into noncoding regions of the genome. An interesting alternative, however, is a sort of genetic typewriter. The genetic code of a three-letter codon could be adapted so that, instead of coding for 20 amino acids and three stop signals, it might specify letters of the alphabet, numbers 0–9, a period stop, and other symbols. Thus the DNA sequence ACTCAAATTAAC would be translated not as the amino acid sequence threonine-glutamine-isoleucine-asparagine, but as the word MINE. With this code, researchers could write their names in the genes of microorganisms, companies could emblazon their logos over genetically engineered crops, and farmers could "brand" the DNA of livestock.

▶ ***Researchers at the University of California were given a small piece of dried muscle from a preserved quagga, an equine species that had been extinct since 1833. They compared purified DNA from this tissue with that of the horse and the zebra, using genetic probes. They found, contrary to previous beliefs, that the quagga was more closely related to the zebra than to the horse. To make the comparison, the researchers had to prepare a library of quagga DNA. In a sense, the quagga animal lives on in this form.***

DNA clues left at the scenes of crimes are increasingly being used as by the police as evidence. The patterns of whorls, arches and lines in fingerprints are different in every individual and can be used as identification. Our DNA sequences are similarly unique. Forensic scientists are now using molecular (or genetic) fingerprinting in crime detection. In cases of rape, for instance, the attacker's sperm can be analyzed genetically using DNA probe technology. A comparison of sequences from the sperm with DNA from blood samples taken from suspects can provide unequivocal proof of guilt or innocence. Blood left on the clothes and tissue found under the fingernails of victims of violent crime can also provide samples for DNA analysis. DNA clues can even take us back to the dawn of civilization. Swedish researchers are attempting to clone the DNA of a mummified Egyptian who lived 2,400 years ago. There is also the prospect that the genetic programs of extinct species can be rescued from preserved tissue.

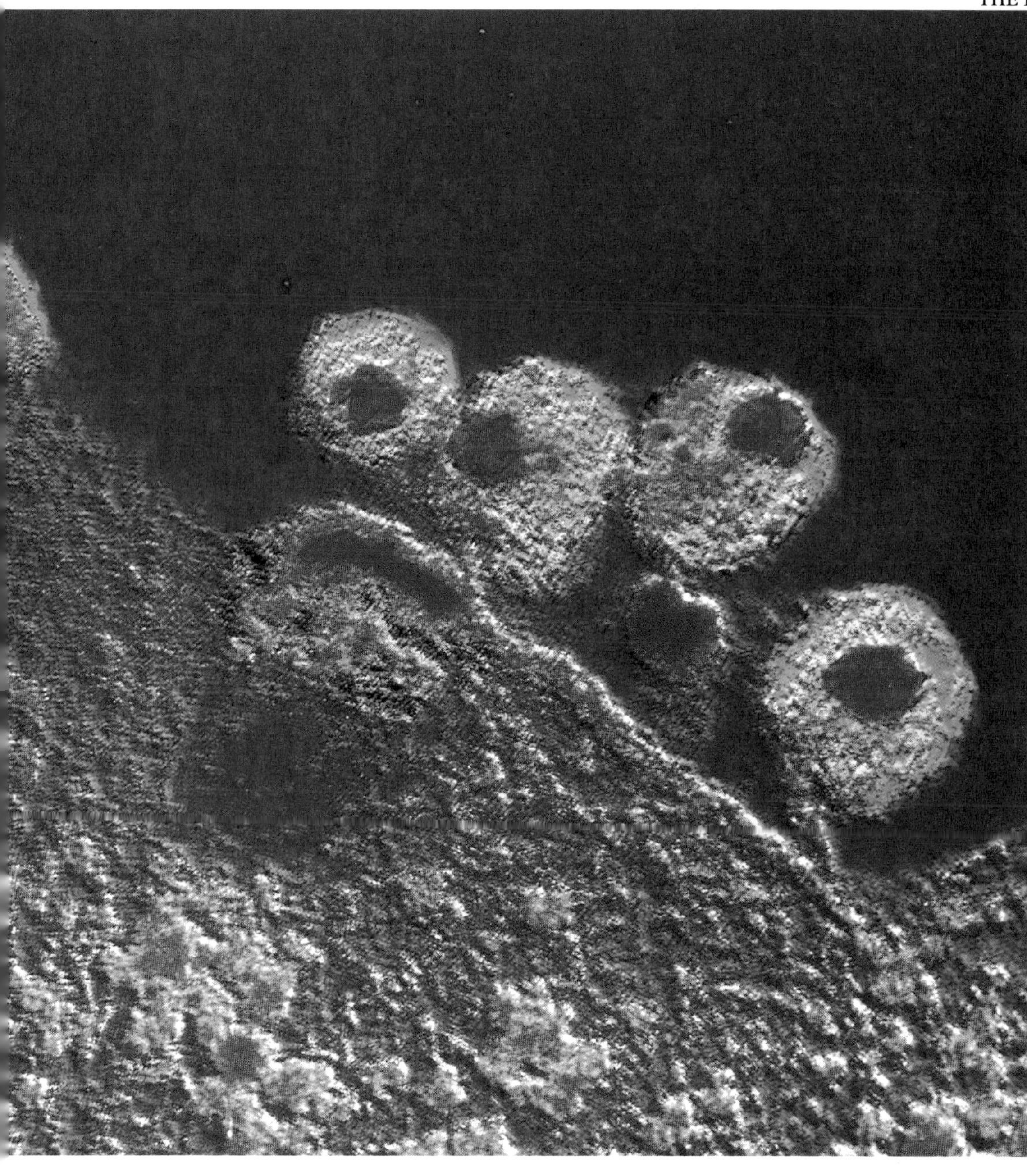

▼ *The ice of Siberia preserved this mammoth leg for over 30,000 years until it was discovered by gold miners in 1986. Nine years earlier a whole baby mammoth, "Dima", had been found similarly preserved. Although the tissue from these specimens is apparently dead, scientists have extracted and analyzed its DNA. This information may help Soviet scientists to regenerate the wooly mammoth. Embryos might be developed from cells of the frozen tissue, and Indian elephants used as surrogate mothers.*

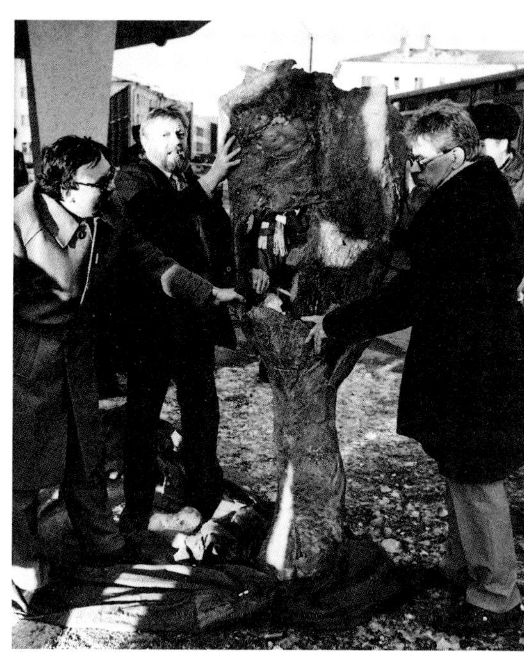

▲ Nerve cells growing on the surface of a silicon chip demonstrate how much smaller is biological circuitry than electronic circuitry. If electronic engineers could use cells, they would be able to design circuits that were thousands of times smaller (and faster) than silicon chips. Miniaturization might be taken further still by taking cellular components from the biochemical circuitry.

Cells as information processors

Computers and living cells are both highly structured processors of digital information. In the former, data and instructions coded in binary form are processed by electrical equipment designed by human beings. In the latter, strings of a base-four code (A, C, G, T) direct biochemical conversions in a chemically-powered machine which has evolved over 3–4 billion years. Unlike computers, however, cells are self-assembling.

The cell is like a complex network of computer modules each connected to a large read-only file-server, the nucleus. This contains an array of software coded as DNA, the individual bases corresponding to digits of information. Discrete parcels of data as messenger RNA are sent out from the nucleus just as binary data circulate in a computer network. In transit to the processors, the RNA signal is amplified by being translated into many copies of a protein molecule. Like the mini-packets of information that travel through computer networks, biological data are addressed to the correct processing modules. These are the cell's organelles. The address is the signal peptide, one end of the protein which acts as a molecular pass key to specific entrances.

The biological processor of the cell would gradually lose the program from its memory unless the software were continuously replaced. Protein turnover makes cellular information processing much more flexible. The cell can alter arrangement and distribution of its components according to its needs. Thus metabolism can be controlled in two ways. Protein production is governed by the needs of the cell as a whole, and the proteins in turn respond to localized changes and adjust the rates of biochemical reactions.

The proteins responsible for biochemical processing are called enzymes. Each enzyme accelerates a particular reaction, processing a particular sort of biochemical information in a specific way. For instance, when our cells are low on sugar, an enzyme breaks down starch into glucose. Enzyme reactions are like electronic components in that the output signal depends on the inputs. Some enzymes act like a biological NOT gate converting substance A to substance B, or B to A, depending on which is available. Enzymes which combine A and B to give substance C could be regarded as AND gates, in that both starting materials must be present in order to produce any output. There are also biochemical OR gates, enzymes which can produce products from either of two substances.

The product, or output, of one reaction may be the input for another, or several others. Linking reactions in this way forms biochemical pathways which are a kind of metabolic circuitry. A set of 10 reactions is needed, for instance, to produce energy for the muscles to work.

Branches occur in metabolic pathways when a substance produced by one enzyme (protein that speeds up reaction) provides the input for two different subsequent reactions. It is often at these branch points that metabolism is regulated. One important mechanism is feedback inhibition. Here the final product of a series of reactions effectively switches off the first enzyme in the pathway. This prevents production of unwanted metabolites. DNA transcription into RNA is also carefully controlled.

Using this complex biochemical circuitry, the organelles work as discrete modules. To make it simpler to understand the complicated workings of the cell, biochemists envisage organelles as black box operations. That is, they only need to consider the inputs and outputs of the organelle without understanding the intricacies of every processing step involved. For many purposes, for example, it is sufficient to envisage the nucleus as something which outputs RNA messages and consequently requires a supply of various RNA precursors.

As in a computer network, the operational units interact through interfaces. In the cell, membranes between organelles are the interfaces. Specific transport proteins in the membranes alllow only selected metabolites in and out. Thus chemical communication between organelles is also controlled. The main purpose of cells is to produce more cells. Cells are, therefore, dedicated, rather than general-purpose, information processors.

The Machinery of Life

3

Composition of proteins...Peptide bonds...Amino acid side-chains...Alpha helices and beta sheets...Enzymes and substrates: structure and function...Modeling proteins: by computer graphics...Protein engineering... Industrial uses of enzymes...Enzymes in solvents... Proteins in the immune system...Hybridoma cells; monoclonal antibodies...Catalytic antibodies

Proteins – from the Greek *proteios* meaning "primary" – are vital building blocks for our bodies and are found in all living organisms. They are familiar as an important part of our diet, and in many countries the protein content of processed food is specified on the can or packet. Nuts, meat, eggs and beans are particularly rich in protein.

But protein is much more than just food. If you imagine the cell as analogous to the body, then DNA would be the brain while proteins would be almost everything else. Like bones, they are important for maintaining structure. Like our sensory organs, they can detect signals. They are involved in communication and in movement. Most important, perhaps, they act like hands in that they make things happen – converting one sort of substance into another, joining small molecules together or breaking down large ones into their components. The activities and functions of a cell or organism depend very much on its proteins. The "survival of the fittest", the basis of natural selection during evolution, depends on organisms having characteristics that are endowed by proteins. Those organisms that fit best in a given environment do so because of their complement of proteins. In a sense, DNA is simply the permanent form of the information contained in proteins.

Protein in the eye and brain

The apparently simple and passive activity of seeing involves, at the molecular level, an extremely intricate series of events. In all of these, proteins play a role. Lysozyme, a protein in our tears, keeps the eyes free from infection. The muscles of the iris, those that move our eyes and our eyelids during blinking, contain thousands of proteins, of which two, actin and myosin, are particularly important. These form complexes of parallel fibers which, powered by biochemical energy, move smoothly over each other during the contraction and relaxation of muscles. Proteins called collagens connect the muscles to other tissues and are important in holding the eye together. The cornea is almost pure collagen.

The protein crystallin is involved in increasing the refractive index of the lens so that it can focus light on the retina at the back of the eye. Rhodopsin, phosphodiesterase and transducin are three proteins in the rods and cones of the retina. They convert the light into biochemical signals that can be interpreted by the brain. Rhodopsin, of which there are around a billion molecules in each of the billion or so rod and cone cells, absorbs the light and alters its structure as it does so. Rhodopsin is actually a combination of a protein – opsin – and retinal, a compound that is closely related to Vitamin A. Every photoexcited rhodopsin molecule can activate several hundred molecules of phosphodiesterase. This process is mediated by the third protein, transducin. Phosphodiesterase in turn alters the electrical charges flowing through the nerve cells attached to the retina and thereby sends a signal to the brain. The processing of these signals is not well understood, but it certainly involves protein activity.

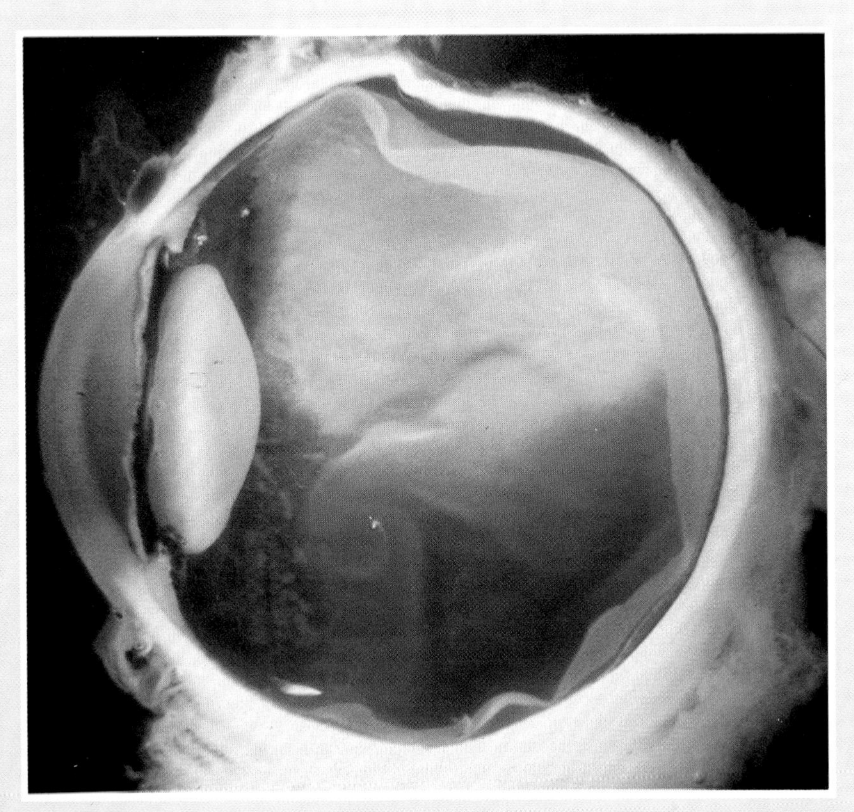

◀ Cross-section of the human eye showing iris, lens and cornea. All its complex molecules are either proteins or are made with the aid of proteins. The names of proteins often end either with "-ase" or with "-in". Thus crystallin in the lens helps the eye to focus. Rhodopsin in rod cells activates phosphodiesterase to send visual signals to the brain.

▲ Enzymes are a class of proteins that promote biochemical reactions. They are usually produced in microorganisms and are widely used in industry to make beer frothy and to make bread rise, to soften leather, to ripen cheese, to produce antibiotics, to clean clothes, and to produce gasoline substitute from sugar cane.

Protein composition and structure

Proteins, like DNA, are linear molecules composed of subunits. They consist largely of chains made from 20 different amino acids (◀ page 10). Every amino acid has three important chemical entities attached to a carbon atom: an amino group, a carboxylic acid group and a side-chain. Chains of several amino acids are called oligopeptides or, if there are more than 20 or 30 amino acids, polypeptides (but the distinction is unclear). Proteins may be just polypeptide chains, often a few to several hundred amino acids in length, but often there are also other chemical groups (such as phosphates or metal ions) making up the protein. In a process called glycosylation, chains of sugar molecules may be attached. Furthermore, several polypeptide chains (either identical or different) may form subunits of a large protein. Hemaglobin, for instance, the protein from red blood cells which transports oxygen, consists of four subunits, each harboring heme, an iron-containing molecule. These cooperate in the way they bind oxygen.

The peptide bond between amino acids is fairly flexible. As a result, oligopeptide and polypeptide chains usually fold into convoluted shapes. Every protein will fold in a particular way to form a distinctive shape suitable for performing its specific function. Since the protein winds round and back on itself, amino acids which are separated by many peptide links in the chain may be very close to each other in the molecule. The way in which the protein folds and operates is dictated primarily by the properties of the amino acid side-chains and their relationships to one another. Some amino acid side-chains are electrically charged (positive or negative). Others – called polar molecules – are neutral but strongly attract electrons. A third group of amino acids have nonpolar, "hydrophobic" (water-hating) side-chains. Charged and polar molecules will strongly tend to seek water while the nonpolar molecules will try to avoid it. Thus proteins in water will fold up so that the polar and charged amino acids are on the outside, with the nonpolar groups buried in the inside of the molecule.

The order of amino acids in a protein chain is known as its primary structure. The three-dimensional shape of a polypeptide is called its tertiary structure. The protein chain folds in on itself to form recognizable modules, common to virtually all proteins. This is the secondary structure. In an alpha helix, for instance, the peptide chain coils round on itself like a loose spring. The coil is stabilized by hydrogen bonds between amino acids four units apart. Another secondary structural element is the beta sheet. Alpha helices and beta sheets provide a further level of stability to protein structures. The secondary structural elements are joined by turns, bends and loops which have varying degrees of flexibility and movement.

▼▶ Beaks and scales, silk and spider's webs are composed mainly of a protein called beta-keratin. In beta-keratin, protein strands are held in position by weak but frequent hydrogen bonds. This arrangement of the protein gives beta-keratin great strength. In a related protein, alpha-keratin, protein strands are coiled in a secondary structure called an alpha helix. In hair and wool, nail and horn, several of these coils intertwine like the strands of a rope.

▶ Each amino acid has an amino group at one end and a carboxyl group at the other. A reaction between the carboxyl group of one amino acid and an amino group of another forms a peptide bond, linking the two together. The resulting molecule, called a dipeptide, also has an amino group at one end and a carboxyl group at the other so that it can form a further peptide bond to build up a protein chain.

▼ *The amino acid cysteine (balls) is important in maintaining the shape of proteins. The side-chain of cysteine contains a reactive sulfur atom (yellow). When the protein folds up into its normal structure, two of these sulfur atoms may come together and react with each other to form a disulfide bridge (–S–S–) between the two parts of the protein chain. Disulfide bridges make certain proteins heat-stable.*

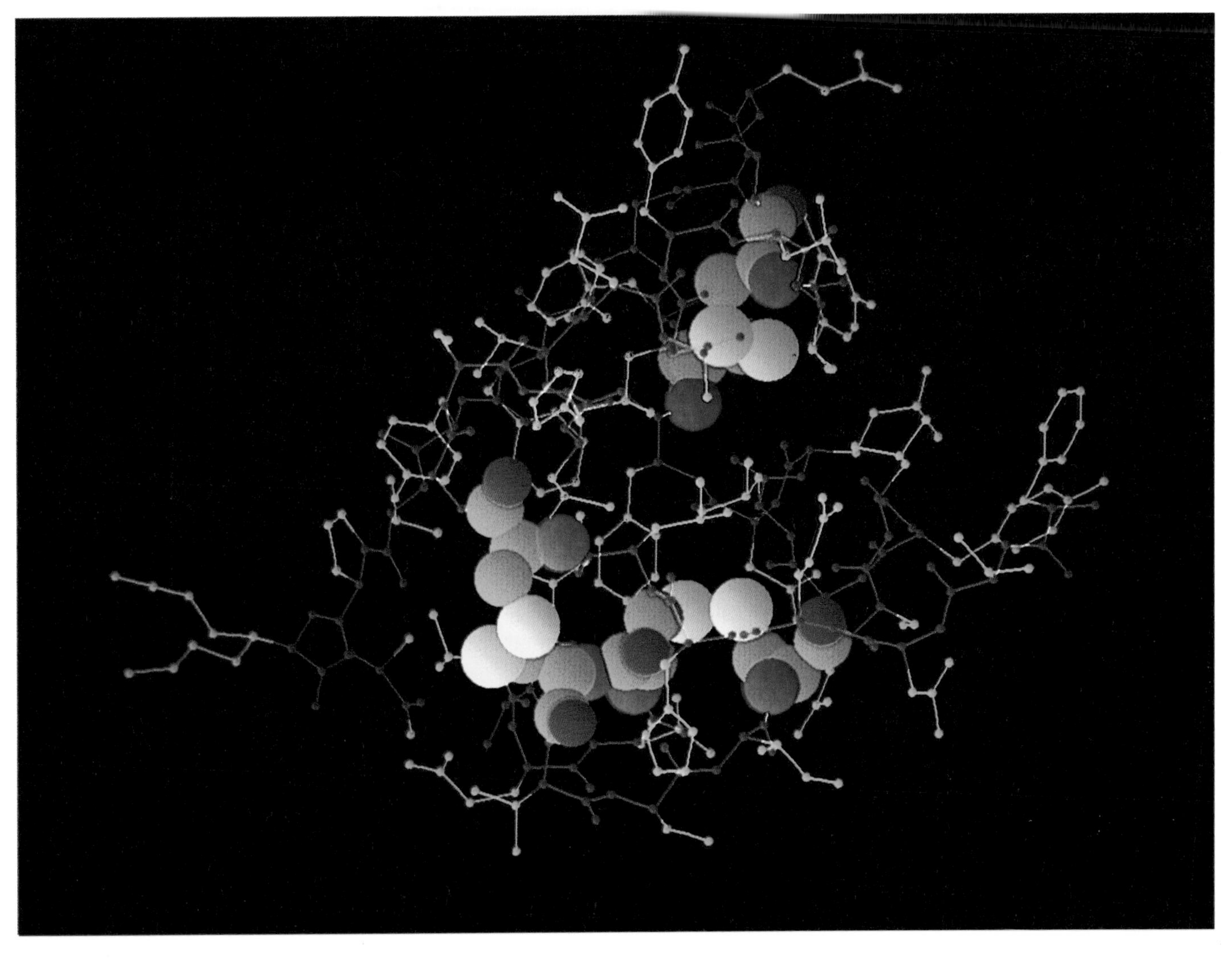

Enzyme action

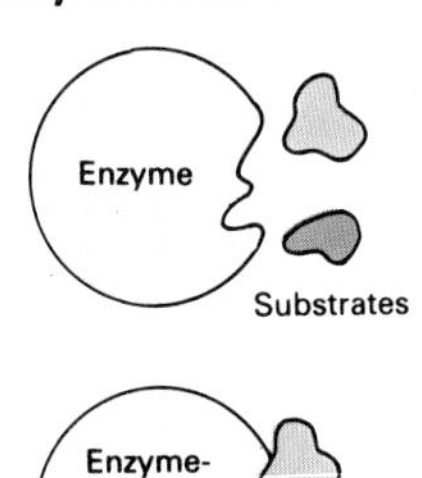

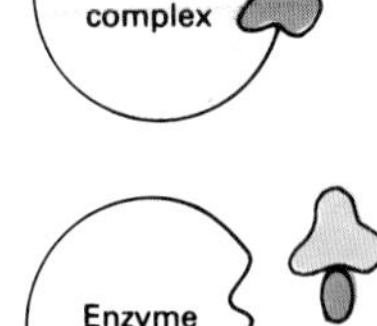

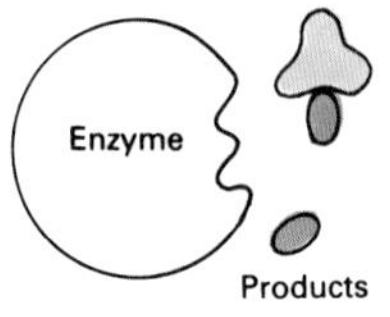

Enzyme inhibition

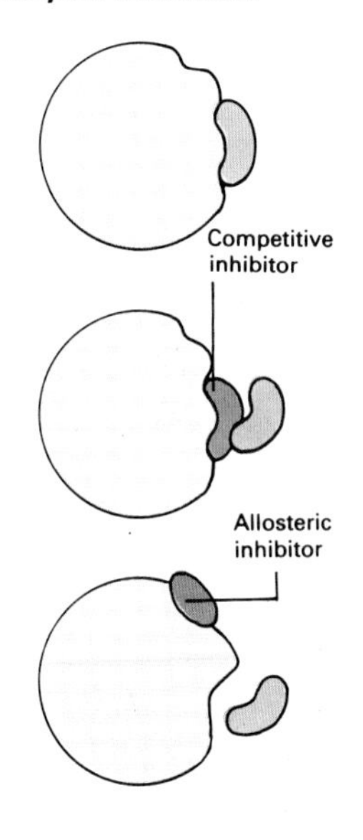

▲ If cells are networks of biochemical plumbing, then enzymes are pumps controlling the flows through the pipes. An enzyme's shape enables it to accelerate a particular biochemical conversion of substrate to product. The pump can be turned off or down by inhibitors. Competitive inhibitors mimic the substrate, blocking the binding site of the enzymes. Allosteric inhibitors alter the shape of the enzyme so it cannot act on its substrate.

Enzymes

Many proteins can speed up biochemical reaction, a property known as catalysis. Such proteins are called enzymes. Thousands of different enzymes are known and their effect is often dramatic. Reactions can proceed millions of times faster in their presence. Enzymes are used in brewing to break down starch into sugar and to clarify beer, in tanning to soften leather, and in baking to yield uniform dough. The active ingredient of rennet, the extract of calf stomach traditionally used to make cheese and junket, is an enzyme called chymosin. In most modern cheese-making, similar enzymes from bacteria and fungi are substituted for chymosin because it is expensive. The enzyme subtilisin appears in many brands of washing powder. It removes biological stains by chewing up proteins. Invertase is used in the confectionery industry for making soft-center sweets. Papain, an enzyme in papaya fruit, helps to make meat more tender.

For commercial purposes, enzymes are produced in tonne quantities either by fermentation or extraction from natural sources. The enzymes used in medicine are usually made in smaller amounts but in a much purer form. For instance, particles containing the enzymes urease and glutamate dehydrogenase are important components of kidney dialysis machines. The enzyme penicillin amidase converts the naturally-occurring form of penicillin into a more active antibiotic. The catalytic power of enzymes has been exploited for a long time.

How an enzyme's structure controls its function

Enzymes are normally much larger than their substrates (the molecules on which they act). For instance, each of the enzymes that synthesize amino acids (and several different enzymes will be needed for each amino acid) itself contains hundreds of amino acids. They have to be large because in each cell of an organism there is a complicated network of metabolism. Thousands of reactions may occur simultaneously, each catalyzed by a different enzyme, and the end result of this activity must be the production of appropriate quantities of biochemical building-blocks and energy to enable the cell or organism to survive and grow.

Since there are thousands of different biochemicals in each cell, many of them very similar, enzymes have to be very specific in their action. An enzyme responsible for joining together two molecules must join only those two and not others. It must not even link one wrong molecule onto one right one. If mistakes were made even in one percent of reactions, chaos would soon reign in the cell. In order to achieve this specificity, enzymes possess a substrate-binding site. This is often a cleft or a hollow formed by the arrangement of the folded protein chain which matches the shape and chemical nature of the substrate to be bound. Biochemists refer to the fit between an enzyme and its substrate as a "lock-and-key" relationship. The substrate fits precisely and exclusively into the binding site of the enzyme.

The enzyme must be specific not only in the substrate that it binds but also in what it produces from that substrate. The specificity of this conversion is due to another functional part of the enzyme, its active site. This is often a small region, sometimes only one amino acid, which is an integral part of the binding site. Where similar reactions are promoted by several different enzymes, the active site in each case can be similar. The rest of the protein chain forms a semi-rigid scaffolding to hold the substrate-binding and catalytic (active) sites of an enzyme in the correct place and in the correct orientation.

Enzyme activation

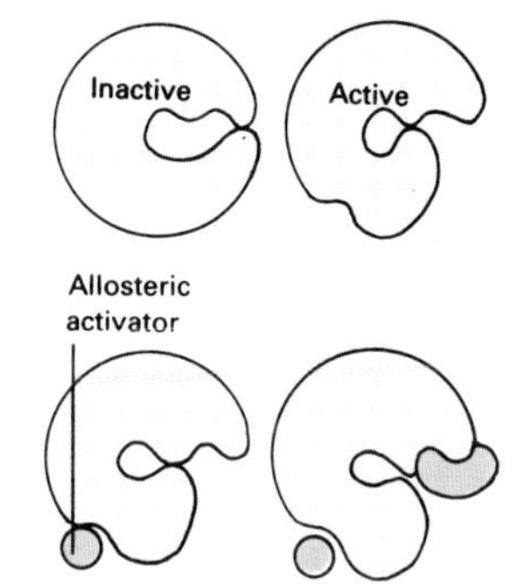

Cooperative effects

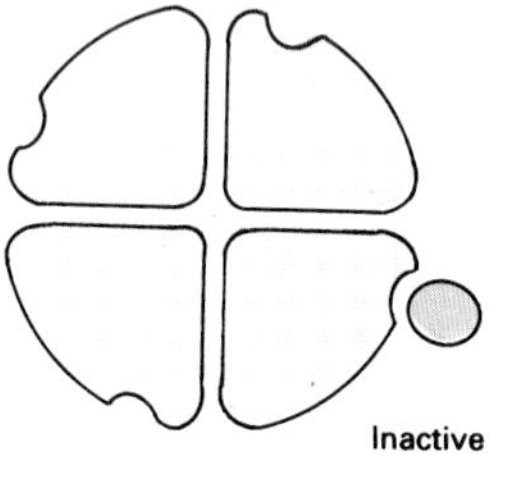

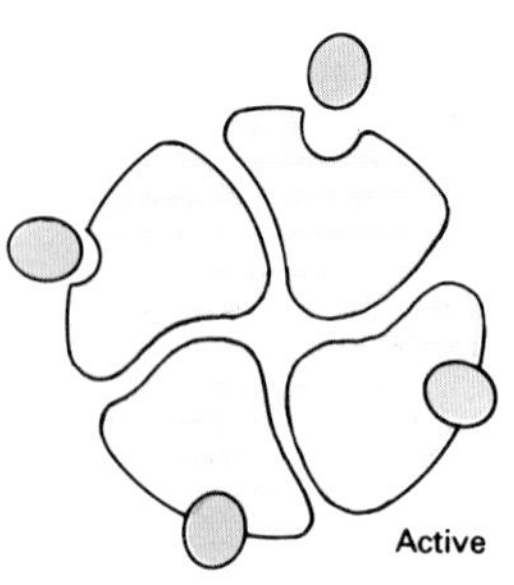

▲ So that cells can react quickly, they sometimes produce inactive enzymes. These become active only when allosteric activators are present to change the shape of the enzyme. This may occur when there is a shortage (or excess) of a key metabolite. In multi-subunit enzymes, a substrate binding to one subunit may activate the others.

► Lysozyme, an enzyme discovered by Sir Alexander Fleming. Experiments with lysozyme have provided much useful information on how enzymes work. Lysozyme has a clearly defined catalytic cleft, a channel that guides its substrates towards its active site.

Enzymes must have stable shapes. Too loose a structure could encourage substrate binding and conversion to be nonspecific. Too stiff a structure, however, would prevent substrates from entering or products from leaving. The reaction rate would thus be slow and difficult to modulate. Heat will tend to make enzymes less stable, while cold will make them less active. The structure of proteins compensates for the environment in which they must operate. The enzymes from microorganisms which grow in hot springs may have additional stabilizing elements, whereas enzymes from Arctic fish have structures that encourage flexibility.

The sequence of the protein chain can also determine whether it is exported from the cell or where it goes within the cell. Information concerning the conditions under which a protein operates, where it has to go, what exactly it has to do, at what speed and under what circumstances, is all stored within the protein's structure.

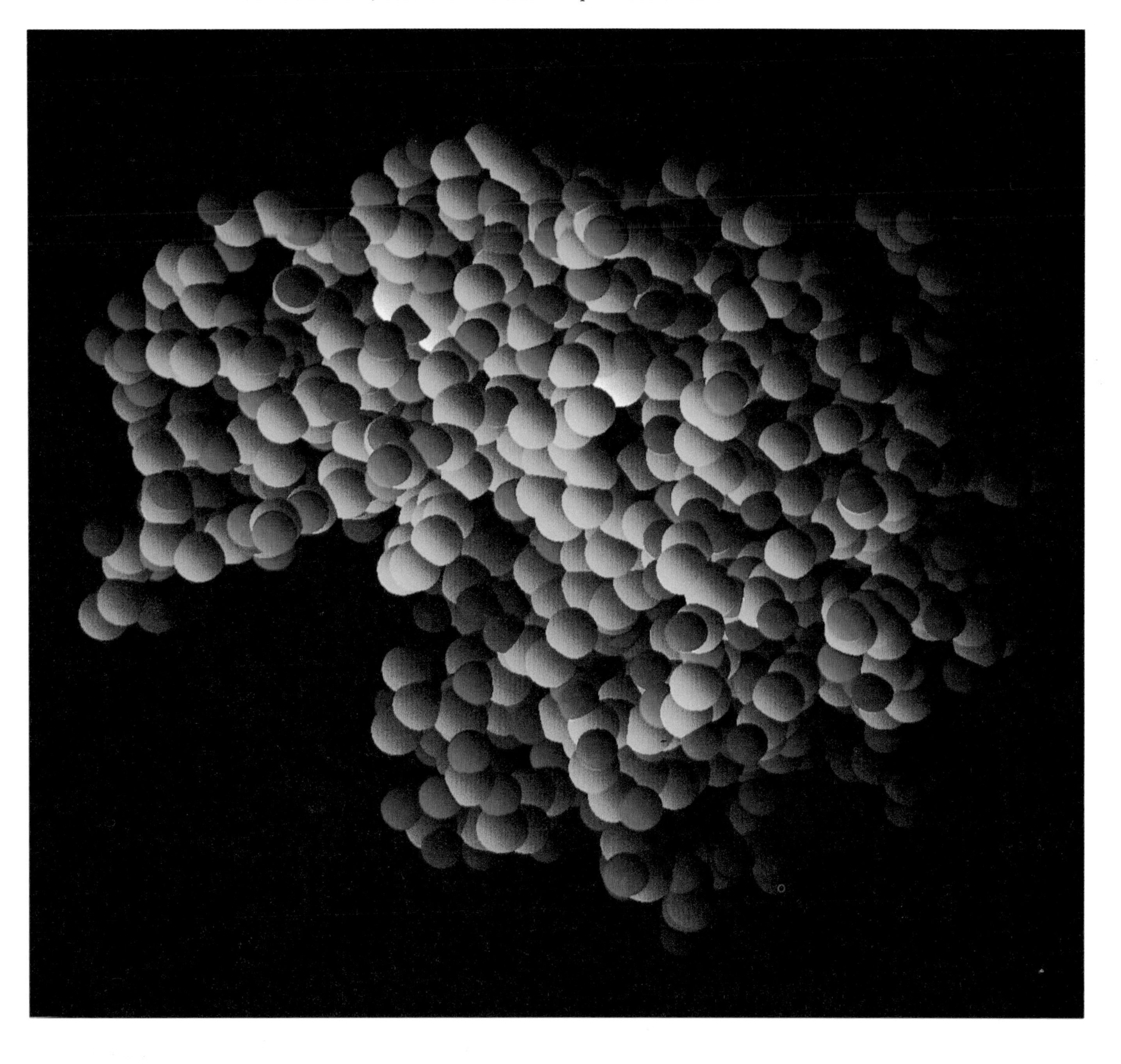

Determining the amino acid sequence of proteins

If a protein can be obtained in a pure form and in sufficient quantity, scientists can fairly easily determine the sequence of amino acids in its chain or chains. In 1967, the first prototype automated protein sequenator was built. It could identify 15 amino acids a day. Modern machines based on much the same technology can "eat" their way through hundreds of residues each day. Sequences of more than 8,000 proteins are stored in computerized form in internationally available databases such as the Protein Identification Resource in Washington, DC. Information on the three-dimensional structure of large biological molecules such as protein is contained in another computerized database, the Protein Data Bank. To define the structure of a protein, biochemists use three-dimensional coordinates. For every atom of a protein, the values of x, y and z define its position in space (relative to an arbitary reference point). To describe the whole molecule (containing several thousands of atoms) information is also needed on which atoms are joined to which others.

▲ *Astronaut Charles Walker checks apparatus for growing crystals of protein. Scientists use X-ray crystallography to analyze the 3-D structure of proteins. Relatively large crystals (0.5mm diameter) are needed to scatter a beam of X-rays into a distinctive pattern of dots. From this pattern, scientists can work out the protein's structure. Several different types of pure proteins, some obtained through genetic engineering were taken aboard the Space Shuttle. Under zero-gravity conditions, the crystals grew much faster than on Earth and were up to 10 times as big.*

Modeling proteins

A model, for scientists, is a representation which allows them to investigate some of the aspects of a real object. Because proteins are so small the models have to be millions of times bigger than the real thing. It is only by studying these models that scientists can gain any idea of how molecules behave at the atomic level.

The instructions for building a skeleton model of a protein are its atomic coordinates as derived from X-ray crystallography. The coordinates are essentially in the form: "There is a nitrogen atom at this position which is attached to a hydrogen atom a fraction of a nanometer over there, a carbon atom up there and another carbon atom there. This in turn is connected ...", and so forth. By scaling up the distances considerably, model proteins can be built, which can be examined from all angles or even picked up. The first ones were constructed of wire, with different-colored wooden or plastic spheres representing the various sorts of atoms. As an alternative representation, interlocking truncated plastic balls can produce "space-filling" models of proteins. The diameters of the balls are determined by the size of the atoms.

► *A model of the oxygen-carrying blood protein, hemoglobin. Working out its structure began in the 1930s and took 25 years to complete. Scientists performed copious calculations without the aid of computers. The model is made of plastic balls and wire, is supported on metal struts, and took several months to build. Altering it in any way is obviously difficult. Today's computer models of hemoglobin (▸ 54–8) can be built in a matter of minutes and changed at the press of a switch.*

Scientists now rarely build models of proteins from plastic kits. They are slow and limited in the information they can represent. Instead, computers are used to generate graphic images of the molecules on a high-resolution screen. They are much faster and can convey many more of the essential qualities of a protein than can rigid models. Atomic coordinate data are in a form that can be readily and directly processed in a computer. The considerable computing power required to produce and manipulate images of proteins indicates how much information is contained in both the physical models and the molecules themselves.

Computer modeling of proteins began in the 1970s, when the calculations for producing the images were performed on mainframe computers. Since then, computing hardware and software have improved so that processing is now performed on minicomputers. The user's interface with the computer consists of a keyboard (for sending instructions to the processor), a regular monitor (which displays the instructions and the processor's messages) and a high-resolution monitor, which displays the molecular image. The screen has around one million pixels, and the image is manipulated by a set of dials.

► Scientists no longer need to work with test-tubes in cluttered laboratories to study the molecules of life. Instead, a computer can display an image of a molecule using data which could have been generated on the other side of the world. With the aid of a keyboard and a set of dials, scientists can examine the molecule from any angle, move it about or zoom in for a close-up view.

Acquiring atomic coordinate data

Information about protein structure is usually obtained by X-ray crystallography. This is a technique in which a beam of X-rays is concentrated on a small crystal of protein which contains 100–200 million million individual protein molecules. Producing crystals of protein is much like growing crystals of salt, but whereas a crystal of salt can be grown overnight by dangling a thread into a strong salt solution, it can take months or years to work out the right conditions for producing a protein crystal.

Gravity causes severe disruption of crystal growth. For this reason, biochemists have tried to grow protein crystals in zero-gravity conditions. Several different types of pure proteins were crystallized in space aboard the US Spacelab and the Space Shuttle. The resulting crystals grew much faster and were up to 10 times as big as those grown under similar conditions on Earth.

When an X-ray beam is projected onto the crystal, it is converted into a distinctive pattern of X-ray dots known as a diffraction lattice. This is recorded using a special camera or an electronic detection system. Different patterns are produced when the beam is projected onto the crystal from different angles.

The diffraction lattice can be converted to an electron density map (like a very complex contour map) of the molecule. Bigger atoms have more electrons and the electron density map indicates both the probable position and size of individual atoms. Scientists use the electron density map, their knowledge of the protein's sequence and the structure of other proteins to deduce a likely structure for the molecule. A computerized simulation then tests whether the proposed structure could have given rise to the original diffraction lattice. Many, perhaps hundreds, of refinements of the proposed structure will be needed before a correct one is found.

To generate a realistic protein model, the computer must transform a flat two-dimensional image into a three-dimensional solid object

Computer graphics are produced entirely by processing numbers. Atomic coordinates provide the framework for the construction of the image and the various atoms are color-coded and sized according to the chemical elements they represent. For a computer to generate realistic images, however, it has to process additional information, adding visual cues to transform a flat, two-dimensional image into a three-dimensional solid object. Light does not pass through solid objects, and so the computer has to calculate which part of the molecule in a particular orientation would be seen and which would be obscured. The farther away the atoms are from the observer, the smaller they must appear. Bonds linking atoms must narrow as they recede. To make the spheres and cylinders representing the atoms and bonds appear solid, the computer calculates how to apply shading and highlights. Light and shade are also manipulated in depth cueing – that is, parts of the object farther from the observer will appear not only smaller, but also darker.

The computer-generated models can depict the properties of amino acids in a way that assists the understanding of how the protein works. Ball and stick models of proteins can be very confusing. Scientists frequently simplify the structure by considering it as a set of different functional elements. For instance, alpha helix and beta sheet secondary structures very often act to "stiffen" a protein (like the stays in a corset) and are represented as tubes and pleated ribbons, respectively. Hydrophobic pockets, small regions of proteins where several water-hating amino acid side-chains act together to hold a

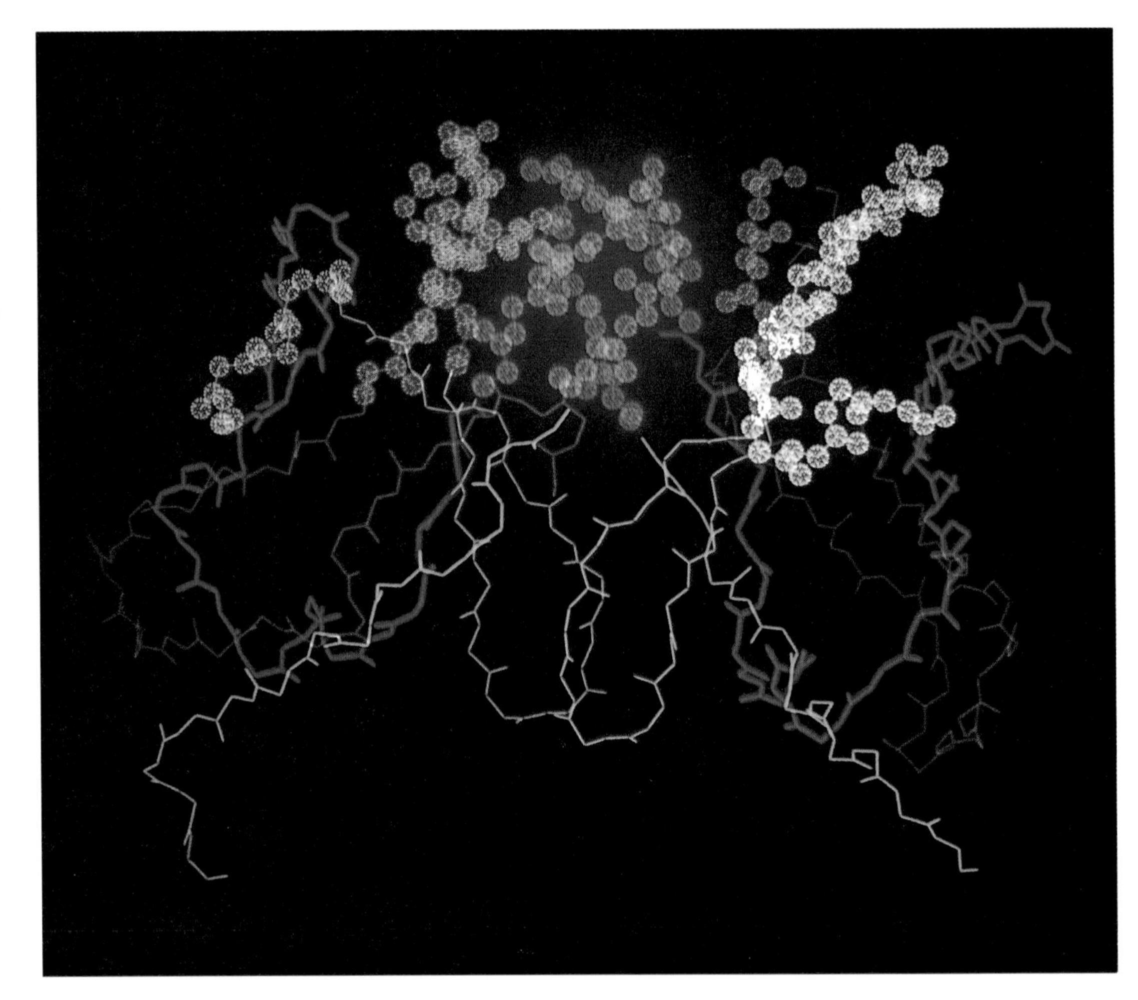

▶ A still from a movie of protein movements. Very high-powered computers work out the position of each atom of the molecule every millionth of a billionth of a second and store the information as graphics. When a sequence of these graphics is played back, the result is a superslow motion cartoon. A cartoon that lasts 15 minutes shows events that actually occupy less than a billionth of a second. Generating the cartoon requires several hundred hours of computational time on a supercomputer.

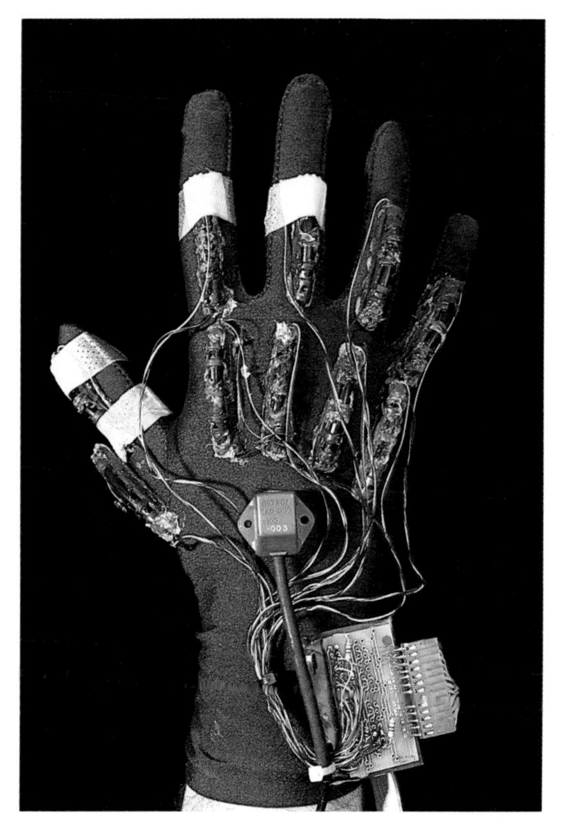

▲ In the future, biochemists may be able not only to "see" proteins, but also to "feel" them. Prototype gloves have been produced which contain electromechanical sensors to interpret the movements of the wearer's hands. If the gloves were connected to a graphics computer, scientists could manipulate the image of the protein just as a sculptor molds clay. With the further development of robotic components added to the gloves, the computer could simulate the shape and feel of the protein and transmit qualities such as texture and resistance to deformation back through the gloves. Such interactive technology, although only in the planning stage, adds new meaning to the phrase "protein modeling".

hydrophobic part of a substrate in place, can be highlighted by color-coding hydrophobic amino acids.

The shape of a protein surface, often depicted as a canopy of dots around the molecule, is important for studying the interaction between two molecules. For instance, part of the surface of the protein *Eco*RI, a restriction enzyme which binds to and cuts DNA, fits beautifully into the grooves of the DNA double helix. Another important feature is the weak electrostatic field surrounding the enzymes, which may draw substrates into the binding and catalytic region. The field is depicted as a series of arrows, giving a clear impression of its purpose.

Current graphical models depict proteins as inflexible structures, but most proteins, especially enzymes, are neither rigid nor static. Rather, they can change their shapes in the same way as a stiff spring can stretch and be compressed, yet return to its original shape. Scientists can estimate how particular parts of a protein will move by considering the strength of its intramolecular interactions. The movements occur extremely rapidly. Shifting a group of atoms from one place to another during enzyme catalysis takes only thousandths, or thousandths of millionths, of a second. But this is regarded as slow since the time taken for molecules to flip into a new position is only a few picoseconds (millionth-millionths of a second).

Movement at the molecular level can be represented as heat. The faster the molecules move, the hotter they are. The mobility of different parts of the protein is color-coded, from dark-gray for the least mobile, through red, orange, yellow and white. This produces a "glowing coal" model of proteins, which is used to predict which fragments of a medically important protein might make the best vaccine. Our immune defences often react most strongly to the "hot spots" of protein "embers".

▲ Using spectacles containing liquid crystal glass, scientists can generate 3-D images of proteins. The computer produces two slightly different images, one for each eye, and flashes them alternately on the screen. The liquid crystal lenses are synchronized so that the left eye sees only the left image and vice versa. The result is convincing 3-D, with the protein apparently floating in space.

Starting with the atomic coordinates of a protein, a computer can build up an image resembling the style of the plastic and wire models

To generate useful models of proteins, computers perform lots of calculations, starting with the atomic coordinates of the protein. These are just dots in space indicating the centers of all the atoms in the molecule (1). The computer could link the points virtually in an infinite number of ways but only one is correct. With the bonds in place (2), the molecule begins to take shape. Adding more detail does not always yield more information. Depicting the atoms as balls (3) confuses rather than clarifies the model. The computer can color-code the different sorts of atoms (4) to enhance the image. But now the computer's depth coding function has been switched off, making the image appear flat. Restoring the differential shading of foreground and background creates an image (5) which closely resembles the style of the plastic and wire models.

1

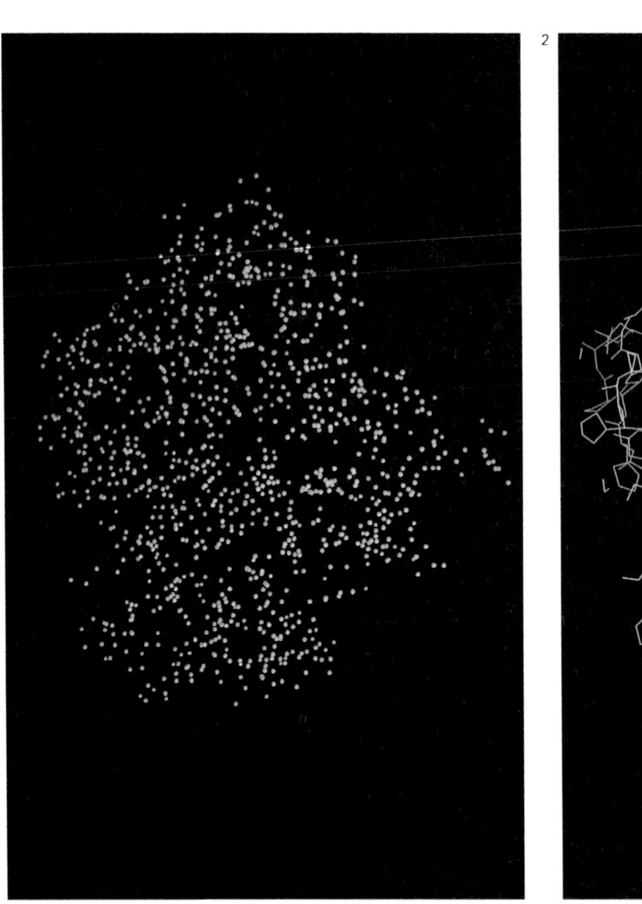

2

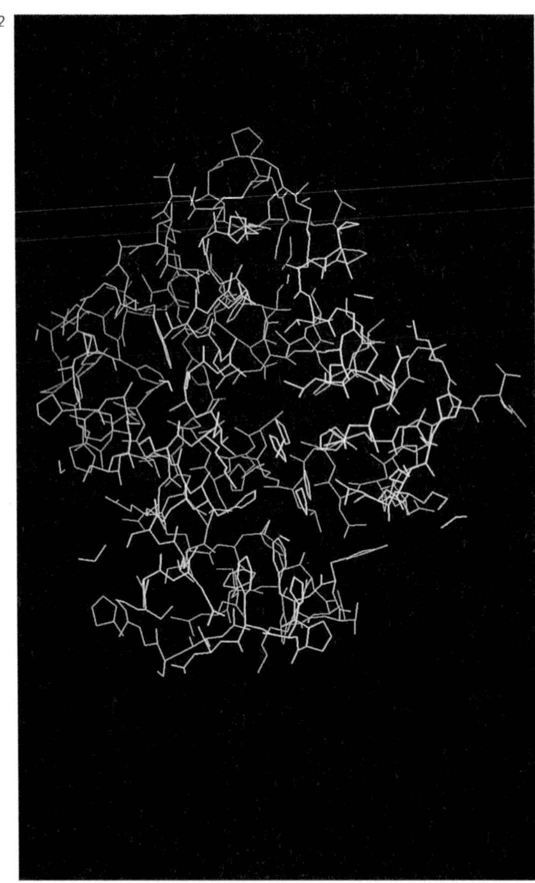

3

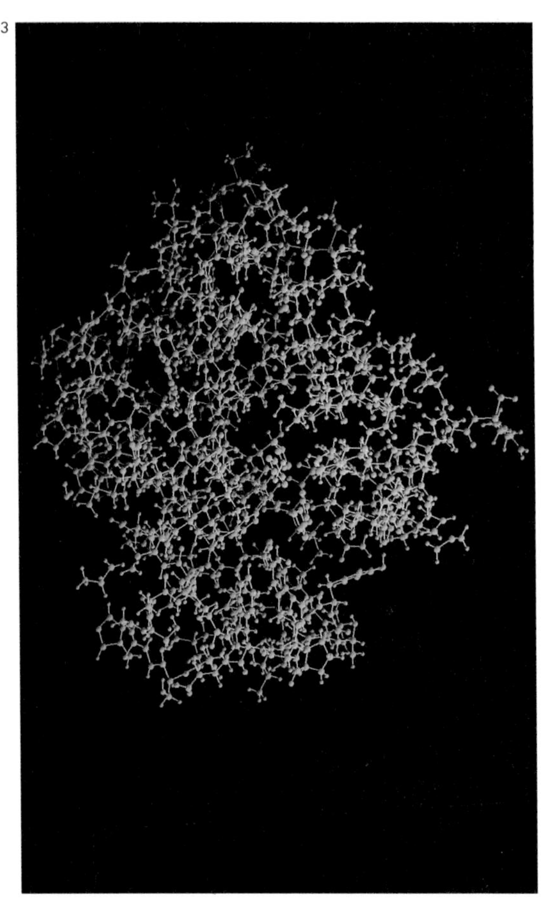

4

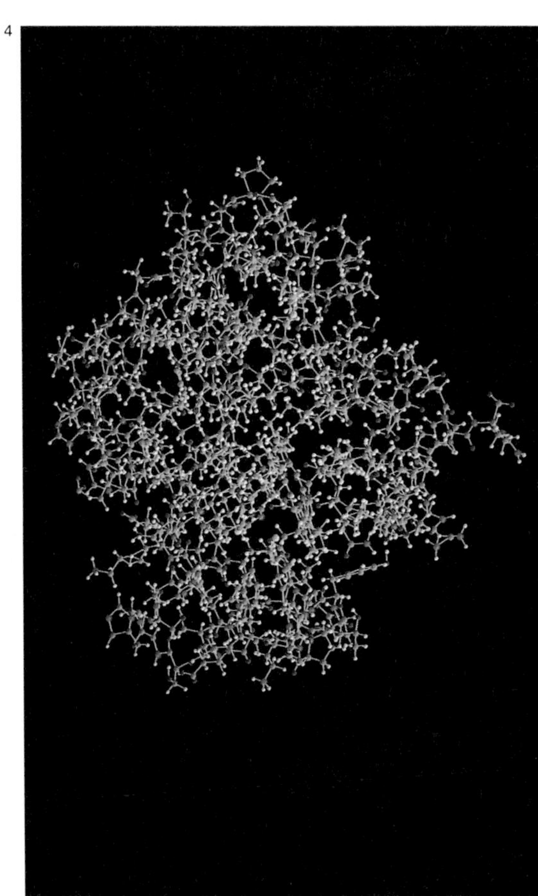

5

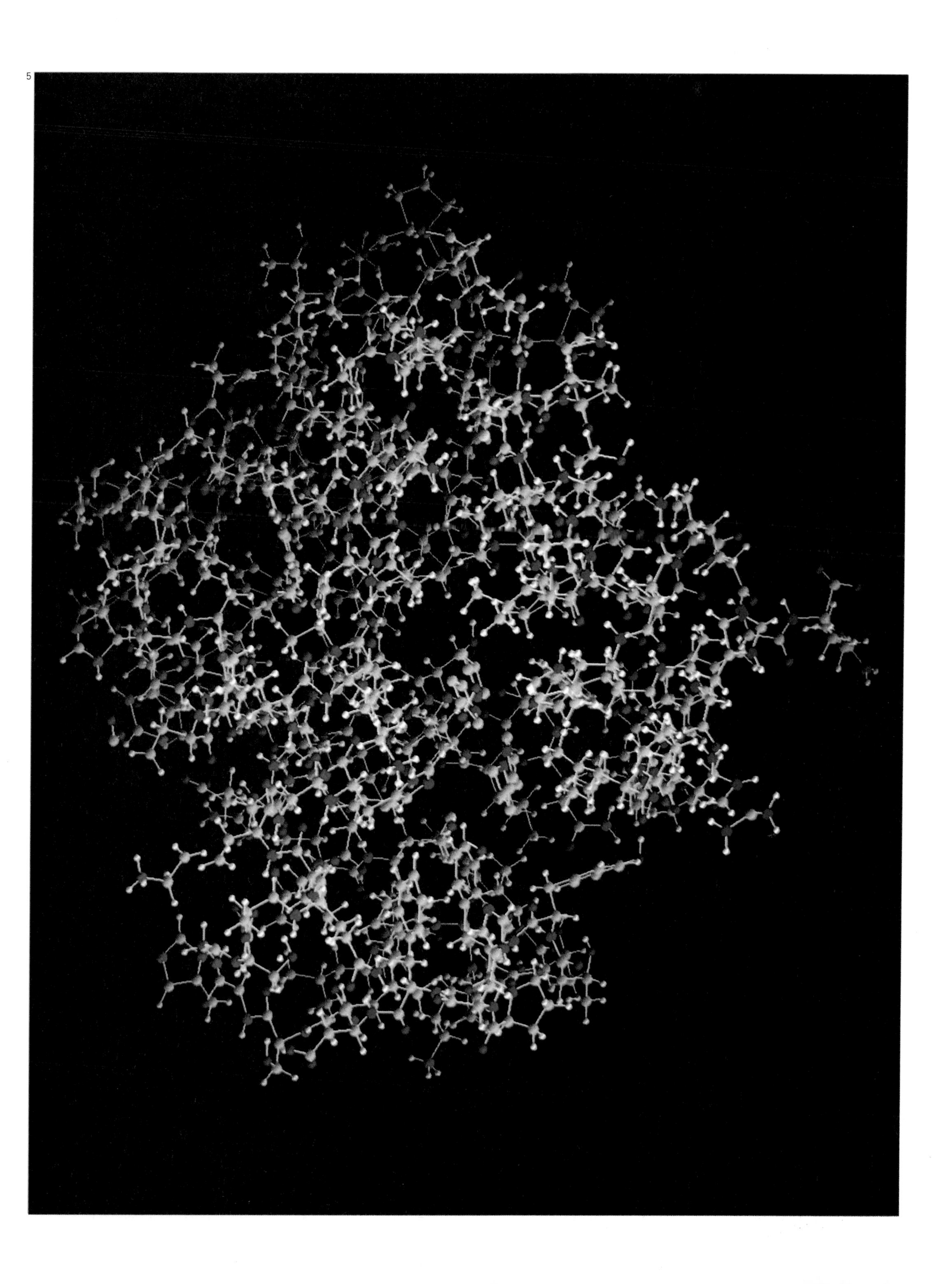

Using a computer to construct a space-filling model of a molecule involves a number of specific stages

The computer can represent hemoglobin as a space-filling model. First, it again converts the atomic coordinates, a set of dots in space, into a set of dots on a flat screen (1). Around each dot, the computer draws a circle whose diameter reflects the size of the various atoms (2). For clarity, the small hydrogen atoms are not shown. Carbon (green), oxygen (red) and nitrogen (blue) are similar in size. The computer then applies standard geometry to shade and transform each circle into a sphere (3). Hemoglobin is actually composed of four very similar protein chains. The computer can bring them together as they would be in the molecule (4), using a different color code for each chain (5). The oxygen-carrying heme complex can also be highlighted (gold). To construct an image of a protein, the computer will perform the various tasks outlined here, though not necessarily in that order.

1

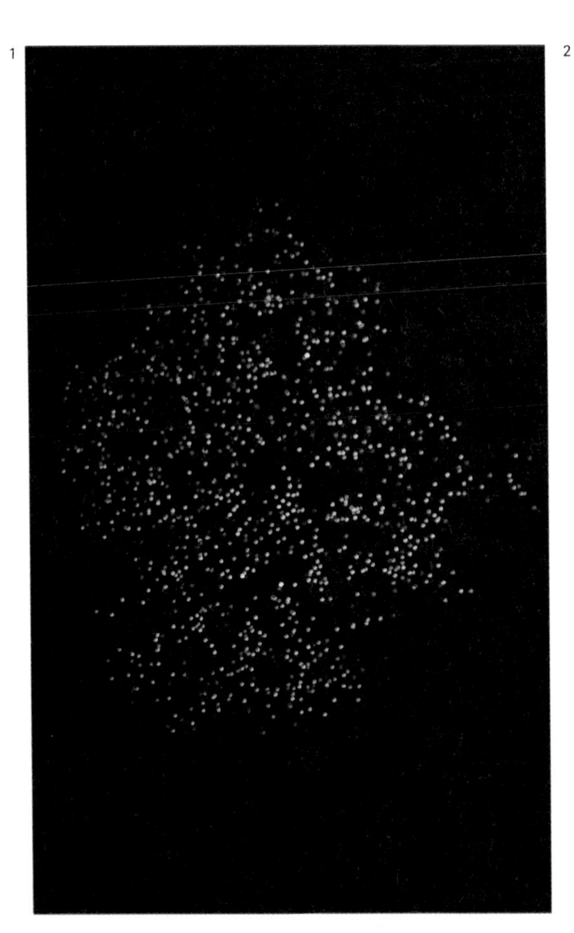

2

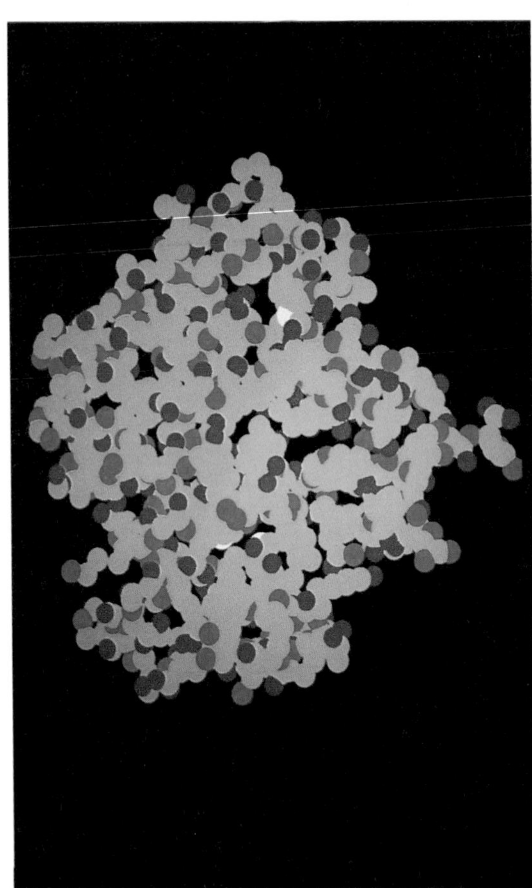

3

4

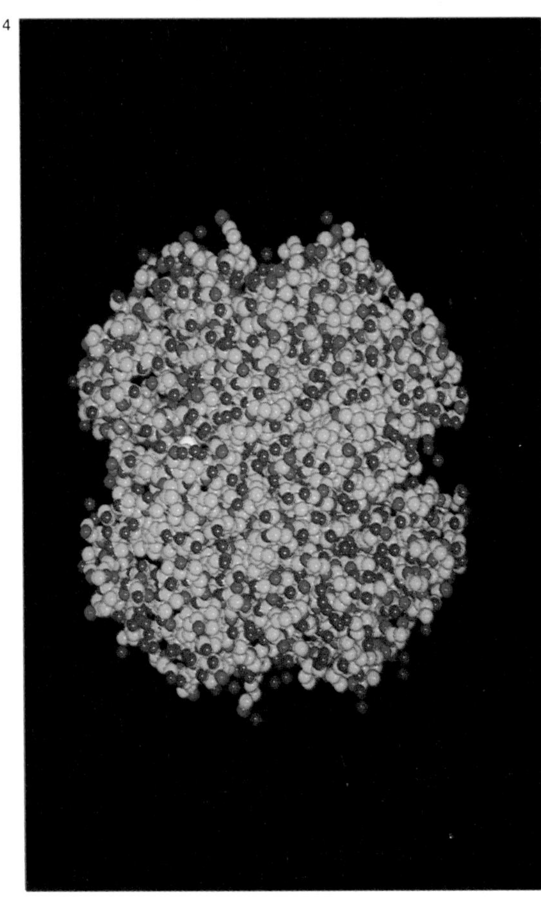

5

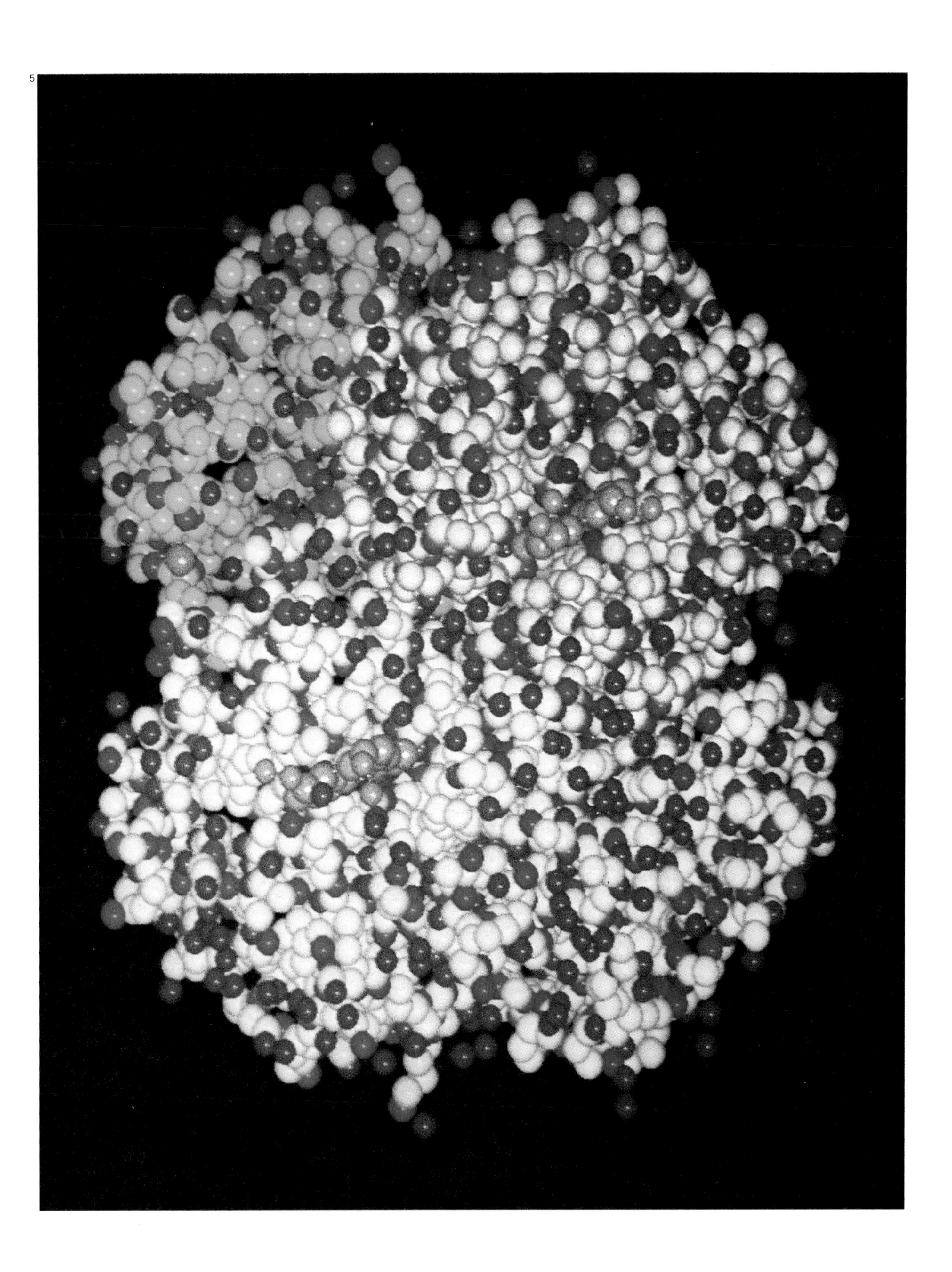

Simplified models of proteins can provide the scientist with an abundance of information

2

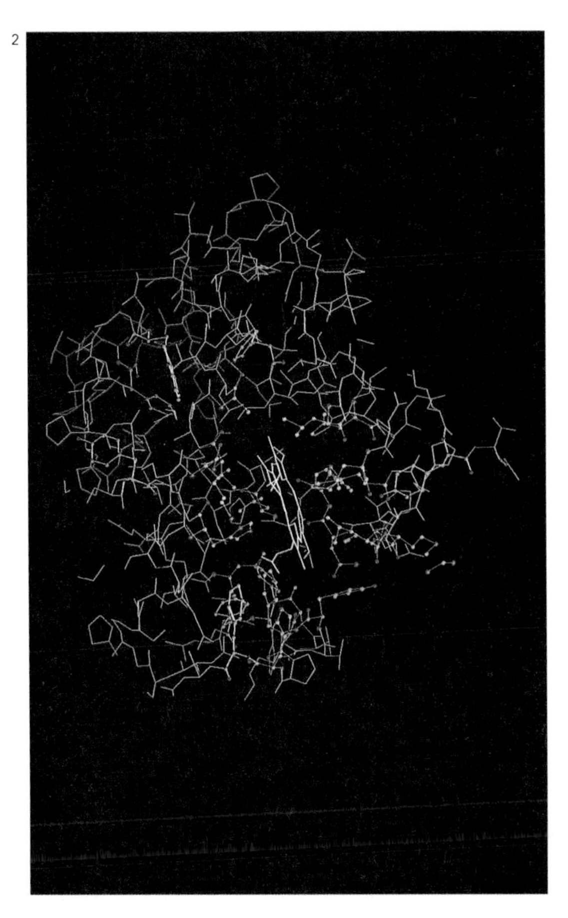

3

The ball and stick and space-filling models are spectacular but often too detailed. A simpler image is obtained if the computer represents alpha helices as cylinders and beta sheets as ribbons. This treatment reveals hemaglobin (1) as a cage of alpha helices surrounding the oxygen-carrying heme group (red). The computer can also identify parts of the protein with particular properties (2). Amino acids touching heme are red. Water-hating amino acids are yellow. The image can also be simplified by zooming in (3 and 4). The computer can then add further details. Arrows (3) indicate the direction of the electric field around part of the heme group while the dot surface (4) provides a transparent space-filling representation. Complicating the image can also sometimes help. A stereo pair, two images at slightly different angles (5), will, with the aid of a special viewer (♦ page 53), appear as one 3-D image.

4

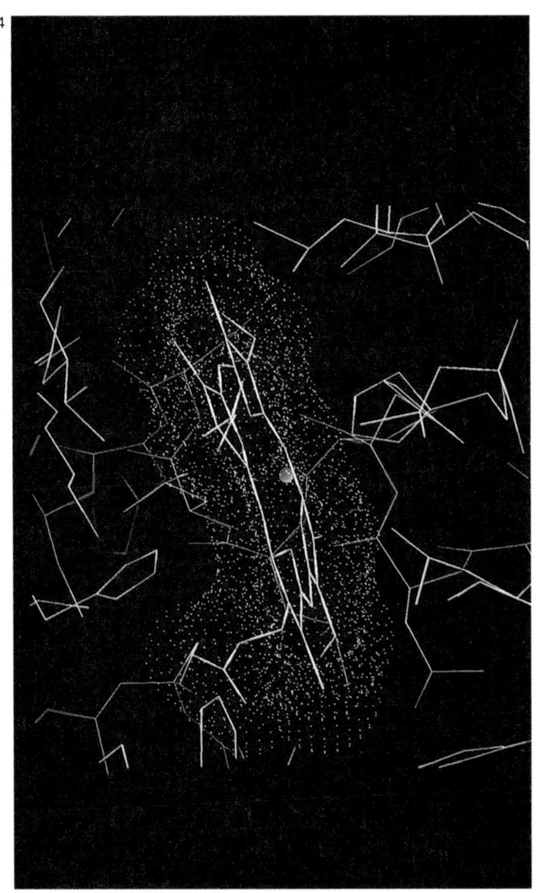

5

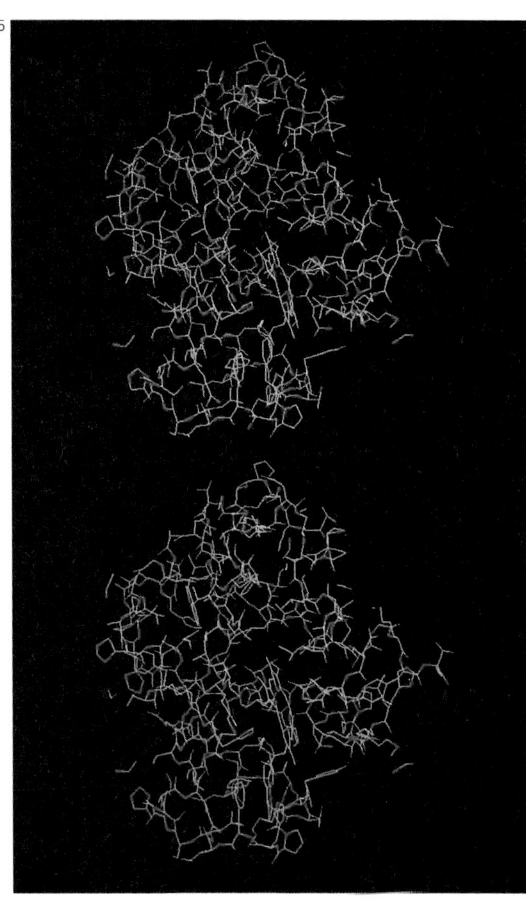

Scientists are not only increasing their understanding of existing proteins, they are beginning to design new ones

Predicting protein structure

How do proteins fold up into the three-dimensional structure that enables them to catalyze a particular biochemical reaction? All the instructions are contained in the amino acid sequence of the protein. When pure preparations of enzymes are heated, they lose their catalytic activity and their three-dimensional structure. When the enzymes are cooled again, both are restored. There is enough intrinsic information in the amino acid sequence to ensure that each molecule adopts its correct conformation.

Scientists' knowledge of the three-dimensional structure of proteins allows them to understand how the molecule works, but only 300 or so protein structures are known. On the other hand, the amino acid sequences of over 8,000 proteins have already been worked out. If scientists knew how proteins folded, they would be able to convert acid amino sequences into structures.

To try to understand protein folding, scientists are adopting a "knowledge-based" approach. They have combined two information databases – one on protein sequences and the other on protein structure. The first step has been to develop a computer program that compares sequences of proteins of known and unknown structure. Homologous proteins, those which differ from each other by only a few amino acids, very probably have similar three-dimensional structures. If the structure of one homolog is known, the structure of the other can be proposed. The database can also usefully compare the sequences of parts of proteins. For instance, the structure of tissue plasminogen activator (tPA), a protein used to treat heart attacks, is not known because tPA has not been crystallized. A three-dimensional model of tPA has been suggested on the basis of sequence similarities with four other proteins of known structure.

The study of protein fragments is the basis of a more general method of predicting structure. Its principle is that certain amino acids, and certain patterns of amino acids, are more likely than others to produce secondary structures such as alpha-helices, beta-sheets and various turns and loops. In the 1950s it was observed that the amino acid proline prevents alpha-helices from forming. Since then scientists have developed a numerical index of all 20 amino acids found in proteins, based on their tendency to form helices. At the top of this table are glutamic acid and alanine, at the bottom, proline and glycine. Similar protocols can predict regions of beta-sheets and turns. Proposed structures can be tested theoretically by a technique called energy minimization which, in effect, examines whether there are undue strains and tensions on the bonds between various atoms in the molecule.

There is a long way to go before structure prediction is reliable. Scientists simply do not understand enough about the properties of amino acids nor can they always recognize the patterns in protein sequences. The fundamental problem is that apparently different sequences can have very similar structures and sequences that appear to be similar have dissimilar conformations.

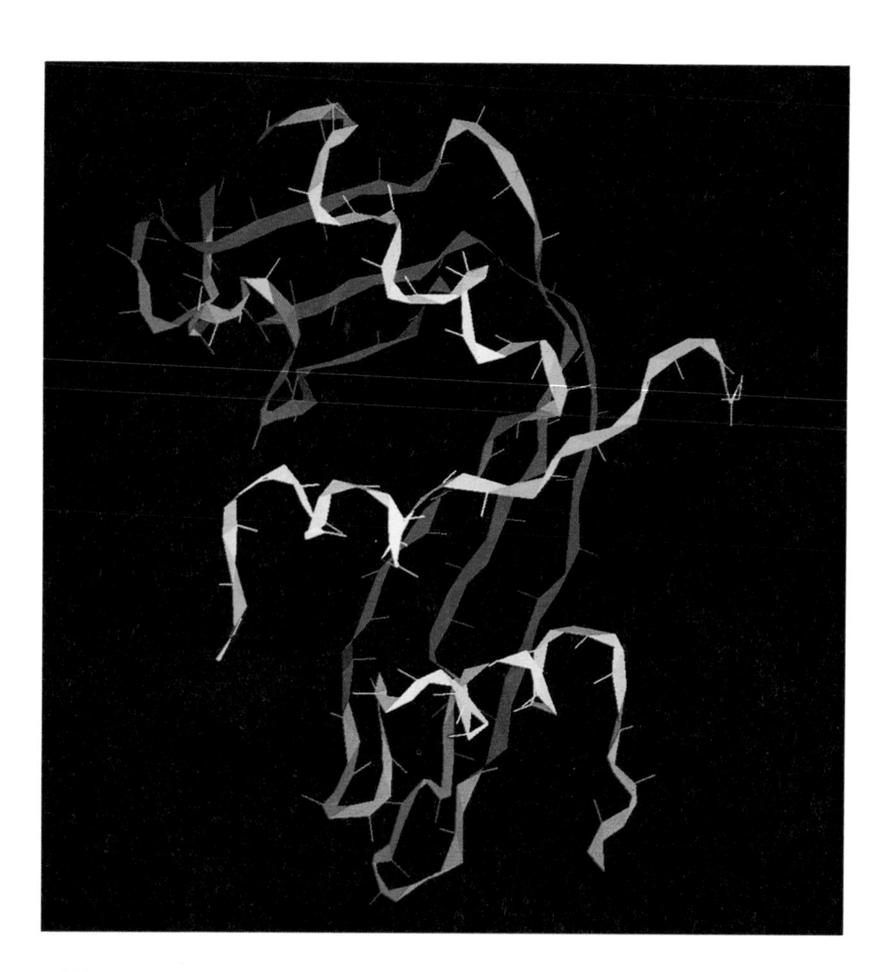

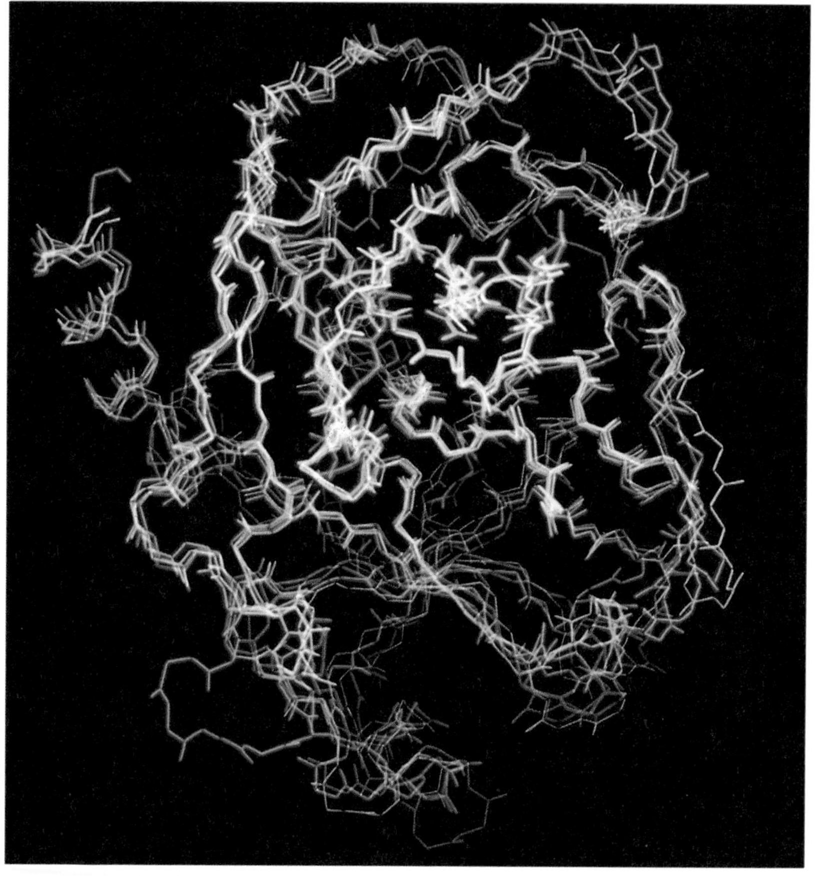

Protein engineering

Scientists are not only increasing their understanding of protein molecules, in some cases they are beginning to produce useful proteins – particulary enzymes – and even to design entirely new ones. Biotechnologists are usually most interested in enzymes that will work hard the whole time. All the complicated feedback control mechanisms built into natural proteins will then be unnecessary. Likewise, the exquisite selectivity of natural enzymes may be wasted when the enzyme encounters just one or a few different chemical types. Indeed, very few proteins, if any, are naturally exposed to the sorts of conditions in which biotechnologists want them to operate. In most cells and organisms, temperatures are ambient, the pH (acidity level) is near neutral, and there are few toxic substances around. Furthermore, proteins in cells have a particular job to do, one specific substrate to convert. Biotechnologists would sometimes like enzymes to act on a range of different substrates.

There is no reason, therefore, why natural proteins should be suited to biotechnological uses. Protein engineers hope to design entirely new molecules. To operate in washing machines, chemical reactors or in food processors, enzymes must survive high temperature, acidity, bleach and organic solvents. Such conditions can damage the proteins, often irreversibly, causing denaturation. This is a familiar process. The proteins in egg white, for instance, are denatured when the egg is boiled. To prevent denaturation, protein engineers endow proteins with new properties by changing any one amino acid in the molecule to any other by a form of genetic engineering known as site-directed mutagenesis. For changes on a large scale, the technique used is cassette mutagenesis. In this, a segment of 20–30 bases of DNA is snipped out and replaced with a segment of the same length but having a different sequence. Often, the DNA inserts are produced in oligonucleotide synthesizers (gene machines), allowing researchers to specify any DNA sequence, and therefore amino acid sequence.

Subtilisin, the enzyme used as an additive in washing powder, has been altered in many ways by protein engineering in attempts to improve its performance. Subtilisin breaks large, insoluble proteins such as those in egg yolk or dried blood into much smaller pieces which dissolve and are removed in the wash. However, bleaching and hot washing, quite normal domestic practices, inactivate subtilisin. Water hotter than about 50°C denatures the enzyme by severing the weak bonds between different parts of the molecule. Scientists at the Genentech Company in San Francisco produced an enzyme which was almost bleach-resistant by replacing methionine amino acid at position 222 of the protein with serine, alanine or leucine. To make the molecule more heat-resistant, the Genentech researchers used computer graphics to identify where different parts of the amino acid chain came in close contact. When they inserted cysteine amino acids at those points, disulphide bonds formed, cross-linking different parts of the molecule. But unfortunately, this did not make the molecule more heat-stable.

Simply knowing the structure of a protein can be of value. Renin is an enzyme produced by the kidneys in response to low blood-pressure. If it is overactive, high blood-pressure can result. But the activity of renin can be reduced by drugs that inhibit the enzyme. Knowing the three-dimensional structures of renin and of the drugs used to slow it down, scientists have been able not only to deduce how the drugs work but also to design improved drugs.

◀ A ribbon model of the protein, ribonuclease showing the various sorts of secondary structure that scientists might be able to predict by examining its amino acid sequence. Alpha-helices (yellow), beta-sheets (red) and various sorts of turns and bends (green and blue) are typical of the major structural elements that make up proteins.

◀ To predict protein structures, it helps scientists to know that proteins with similar functions often have similar structures. The structures of four enzymes (trypsin, kallikrein, chymotrypsin and elastase) which cut up proteins are shown superimposed. Large regions of the amino acids coincide almost exactly, with few dissimilarities.

▲ One of the largest markets for enzymes is in biological washing powders, where they help remove protein-based stains like blood and egg yolk. The search for the ever whiter wash has led scientists to try and improve enzymes through protein engineering. The two main aims are to make the enzymes bleach-resistant and heat-stable.

Enzymes in the chemical industry

Oil, coal and natural gas are mainly composed of organic chemicals. Through the chemical industry, these are converted into many of the substances we use in our daily lives. The products of the fine chemical industry, such as flavors, perfumes and drugs, are usually very complex and require involved chemical syntheses. Increasingly, enzymes are being used to catalyze some of the intricate steps in organic syntheses.

The drug cortisone, which reduces swelling of the joints and pain in arthritis patients, used to be produced by a complicated chemical process involving 37 steps. Only 0.02 percent of the starting material was converted into the final product, and as a result, the drug cost over $200 per gram. Then it was discovered that an enzyme from the mold *Rhizopus arrhizus* could alter the starting materials to cut out many of the chemical steps. A new, much simpler synthetic method was devised and the cost of cortisone fell to below 70 cents a gram.

▼ *The ghostly glows of a large chemical plant. Chemical plants operate at very high temperatures and pressures, conditions not normally favoring enzymes. Enzymes are, however, used under mild conditions to produce certain highly valuable chemicals. In the future, enzymes from "black smoker" bacteria found in very hot, pressurized undersea vents may be used by the chemical industry.*

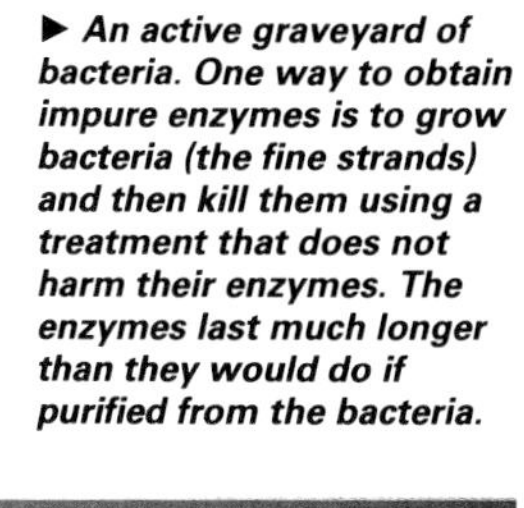

► *An active graveyard of bacteria. One way to obtain impure enzymes is to grow bacteria (the fine strands) and then kill them using a treatment that does not harm their enzymes. The enzymes last much longer than they would do if purified from the bacteria.*

▼ *Protecting enzymes by immobilization. Enzymes dissolved in water efficiently convert substrates to products. However, separating the enzyme from the products afterwards is difficult and expensive and may damage the enzymes. For many industrial uses, enzymes are attached to the surfaces of, or contained within, inert beads (lower picture). In this form, they are robust, and can be separated from the products by sieving or the action of gravity.*

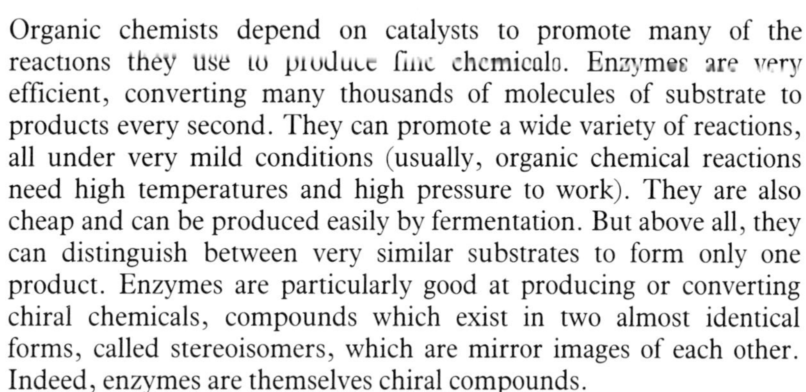

Organic chemists depend on catalysts to promote many of the reactions they use to produce fine chemicals. Enzymes are very efficient, converting many thousands of molecules of substrate to products every second. They can promote a wide variety of reactions, all under very mild conditions (usually, organic chemical reactions need high temperatures and high pressure to work). They are also cheap and can be produced easily by fermentation. But above all, they can distinguish between very similar substrates to form only one product. Enzymes are particularly good at producing or converting chiral chemicals, compounds which exist in two almost identical forms, called stereoisomers, which are mirror images of each other. Indeed, enzymes are themselves chiral compounds.

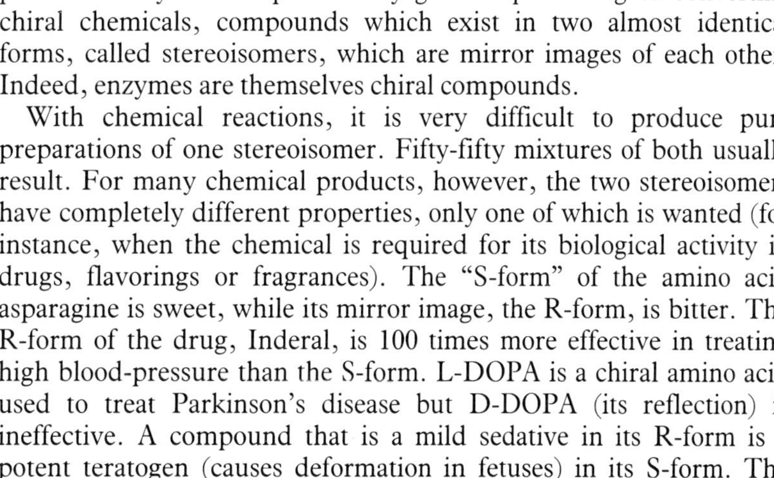

With chemical reactions, it is very difficult to produce pure preparations of one stereoisomer. Fifty-fifty mixtures of both usually result. For many chemical products, however, the two stereoisomers have completely different properties, only one of which is wanted (for instance, when the chemical is required for its biological activity in drugs, flavorings or fragrances). The "S-form" of the amino acid asparagine is sweet, while its mirror image, the R-form, is bitter. The R-form of the drug, Inderal, is 100 times more effective in treating high blood-pressure than the S-form. L-DOPA is a chiral amino acid used to treat Parkinson's disease but D-DOPA (its reflection) is ineffective. A compound that is a mild sedative in its R-form is a potent teratogen (causes deformation in fetuses) in its S-form. The name of the sedative is thalidomide. It is often extremely important, therefore, to produce preparations containing just one stereoisomer.

Enzyme synthetic methods have been of commercial interest since the Nutrasweet Company in the USA started selling large amounts of aspartame, the dipeptide sweetener. Half of aspartame is an amino acid called L-phenylalanine, which rival companies have attempted to produce in quantity using enzymes. Three methods of production have been tried, each using a different enzyme. The diversity of enzyme catalysis can help to overcome many of the difficulties of synthesis faced by organic chemists.

Enzyme activity in solvents

The efficiency, selectivity and specificity of enzymes can be of great value to organic chemists. Organic chemical reactions, however, were traditionally conducted in solvents, such as oil, chloroform or hexane, because most organic chemicals did not dissolve very well in water. Enzymes, on the other hand, are found naturally in living organisms that are mainly composed of water and do not survive in organic solvents. Even a mild solvent like alcohol will deactivate enzymes. Until the 1980s scientists believed that enzymes could not catalyze reactions in organic solvents. Researchers have since discovered that in fact, enzymes often work much better in organic solvents.

One of the tricks used by biotechnologists is to protect the enzyme inside "reverse micelles". A reverse micelle is a small droplet of water in a "sea" of organic solvent. To produce enzymes in reverse micelles, biotechnologists simply make a very oily dressing by adding a small volume of water and a pinch of enzyme to the solvent, and then shaking thoroughly. Inside the reverse micelle, enzymes are surrounded by a protective pool of water. The organic substrate can be added to the solvent in large quantities, with no effect on the enzyme. A certain amount of the added substrate passes into the reverse micelle by a process called diffusion and is converted to the desired product by the enzyme. Conversion reduces the amount of substrate in the reverse micelle, allowing more to diffuse in. This too is turned into product, and so the process continues. If the product dissolves well in water, the reverse micelle will eventually become overloaded with product and the conversion will stop. If the product dissolves better in the organic solvent, however, it will leave the reverse micelle as soon as it is produced by the enzyme. The enzyme will continue to work as long as the substrate diffuses into the reverse micelle, so that virtually all of the substrate ends up as product.

Micelle

▲ A reverse micelle is a droplet of water surrounded by a "skin" of polar lipids in a sea of organic solvent. Enzymes remain active in the water but can act on substances which are not very water-soluble when these diffuse across the "skin".

► Agricultural herbicides are just one of the classes of bulk chemical compounds that could be produced using enzymes. Enzymes are usually found in watery surroundings in cells and organisms. Chemists and chemical engineers are used to working with large-scale reaction vessels containing organic solvents. Thus if researchers are trying to develop enzyme systems which function in organic solvents, some minor modifications to existing production techniques will be necessary.

Apparently, then, so long as enzymes have a little water they will function very well in organic solvents. Scientists in the USSR started to ask just how much water enzymes needed. When they mixed a thoroughly dried preparation of an enzyme into a pure organic solvent, much to their surprise, the enzyme exhibited considerable activity with no water present. (In fact, later research showed that a "skin" of water molecules on the enzyme surface was needed for activity.)

Interestingly, not only could enzymes function in pure solvents, but they catalyzed entirely different reactions there. For instance, lipase enzymes in water break down fats (lipids) into small pieces. In organic solvents, however, lipases reconstruct fats from smaller molecules. By choosing which molecules to supply to the lipases, biotechnologists can determine what sort of fat is produced.

The unusual behavior of enzymes in organic solvents is due simply to the absence of water. Water is not just the liquid in which enzymes float, it also takes part in many reactions. The breakdown of fats, proteins, DNA and other large biological molecules occurs by a process called hydrolysis (splitting with water). Hydrolysis is usually catalyzed by enzymes (lipases, proteases and DNAases) But when water is absent, these splitting reactions cannot take place. Instead, the enzymes catalyze linking reactions, the reverse of their normal activity. Scientists in Japan and Denmark have used proteases in organic solvents to join together amino acids and peptides to make proteins artificially. Enzymes in organic solvents are less easily denatured by heating than in water, and remain stable and active above 100°C.

► The starting material for chocolate-making is cocoa butter, a rich mixture of fats which provides much of the taste and texture of the confection. Cocoa butter is expensive, but scientists are developing enzymic methods of manufacturing a substitute from fats such as palm or rapeseed oil which can be produced more cheaply.

▼ *A pilot-scale reactor at Unilever in the UK which interconverts various types of fat enzymically. Inside, enzymes rearrange the fatty acids that compose fats and oils. This can change many of the fats' properties. That fats are becoming more interconvertible has important implications for cash-crop economies.*

The immune system

Biotechnologists would dearly like to be able to produce proteins in large amounts and with activities that can be altered to order. In fact, there is a natural protein-production machine that almost meets those requirements. It will produce, in large amounts, proteins that bind very specifically to any desired molecule. The machine is part of the immune system called the B cell, and the specific binding proteins are called antibodies.

The purpose of the immune system is to repulse invasions, usually by harmful infectious organisms. Many different kinds of cells of the blood (the white blood corpuscles) and the lymphatic system help to defend the body. The cells are assisted by antibodies which act as molecular adaptors. The two arms of the antibody's Y-shape bind particular features on the surface of invading organisms while the trunk attracts the cells of the immune system. Antibodies enable the immune system to react to almost any invader. If the body is invaded, for instance, by the measles virus, the B cells produce antibodies which bind specifically to the virus and activate the immune cells to deal with it. If the intruder is the *Streptococcus* bacterium, which causes sore throats, they generate anti-*Streptococcus* antibodies, and so on. Anything that provokes B cells to produce antibodies is called an antigen (antibody generator).

Antibodies will react to and bind antigens even if they are not on the surface of invading organisms. This is why pure proteins or carbohydrates extracted from organisms can act as vaccines (➧ page 91). Indeed laboratory animals can even produce an immune reaction against completely unnatural molecules synthesized in chemical laboratories.

The immune system does not "know" which antibody to produce to bind a particular antigen, but it finds a suitable one by a trial-and-error process called clonal selection. When an invader enters the body, there are already B cells which can respond to a degree. When the antibodies from these cells bind the antigen, the B cells are stimulated to multiply rapidly, to produce clones of daughter B cells. During this burst of activity, mutation, reassortment and joining inaccuracies produce variants of the antibody originally selected, some of which show improved antigen binding. The process may be repeated on the cells that produce the new antibody. Repeated clonal selection produces large numbers of identical B cells each making an antibody that specifically binds the invading antigen.

► When an invader (antigen) enters (1) the bloodstream, it may be taken by a macrophage cell (2) to a lymph node (3) where it is matched to a B cell. This stimulates the B cell, aided by other immune cells, to differentiate and multiply into a clone of plasma cells each of which produces huge amounts of antibody. The antibodies flood into the bloodstream (4) to bind to and neutralize the invader.

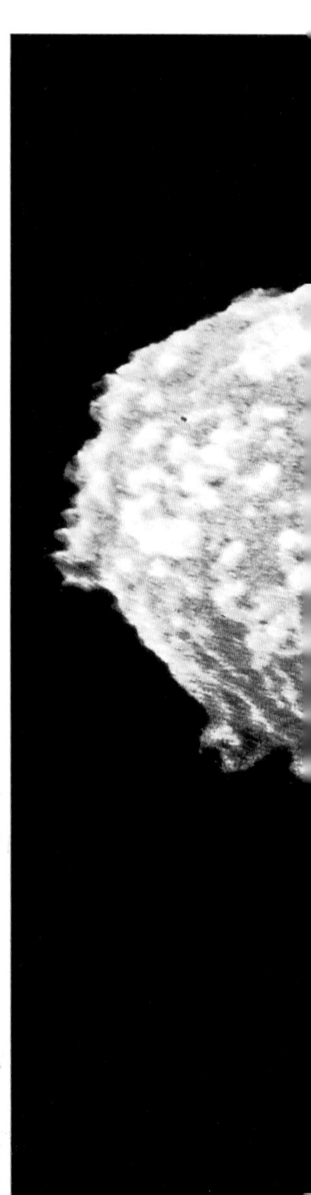

► Antibodies are produced by B lymphocytes, a subset of the white blood cells. These cells form in the spleen and bone marrow and are distinguished from other white cells by the presence of antibodies on their surfaces. A B cell only produces antibodies that bind to one specific antigen. If a B cell is fused to a cancer cell, the result is a hybridoma cell which not only grows indefinitely in artificial culture but also continues to produce the same type of antibody as produced by the B cell.

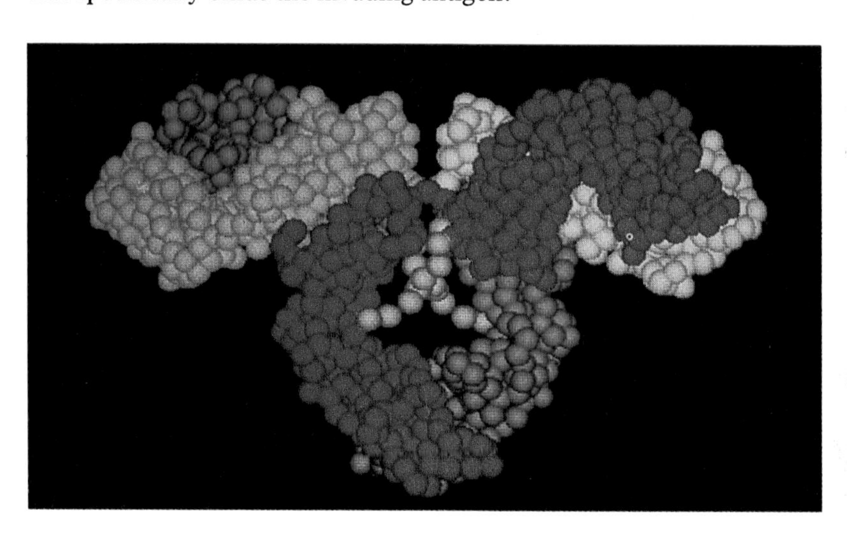

► Antibodies are universal adaptors. Four proteins – two heavy (red and blue) and two light (yellow and green) chains are linked to produce roughly Y-shaped molecules. The trunk of the Y is the "constant" region of the antibody. Its arms are the "variable" regions, responsible for binding to antigens. As many as seven different genes code for different parts of the antibody, providing a great deal of molecular diversity. Frequent mutations in the genes for the antigen-binding regions mean that antibodies that bind to almost anything can be produced.

1

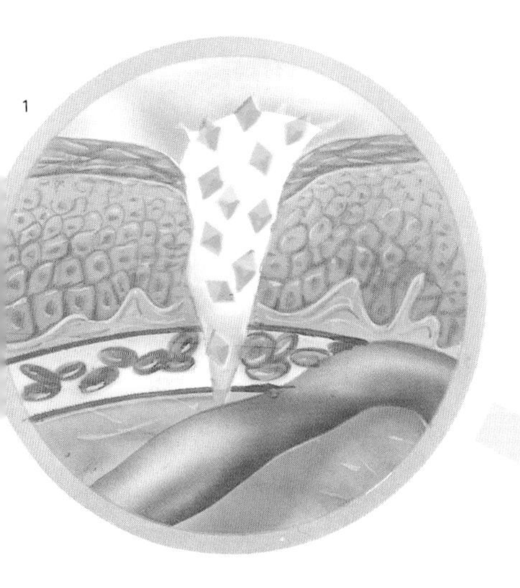

Antigen
Antibodies

Macrophage

Neutrophil
Complement

B-lymphocyte
T-lymphocyte

2

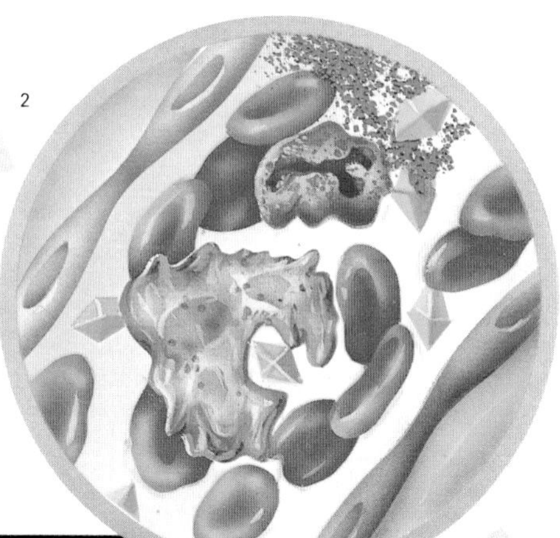

3

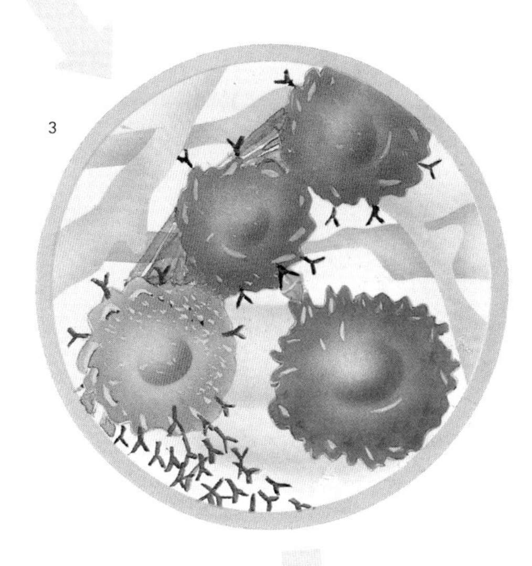

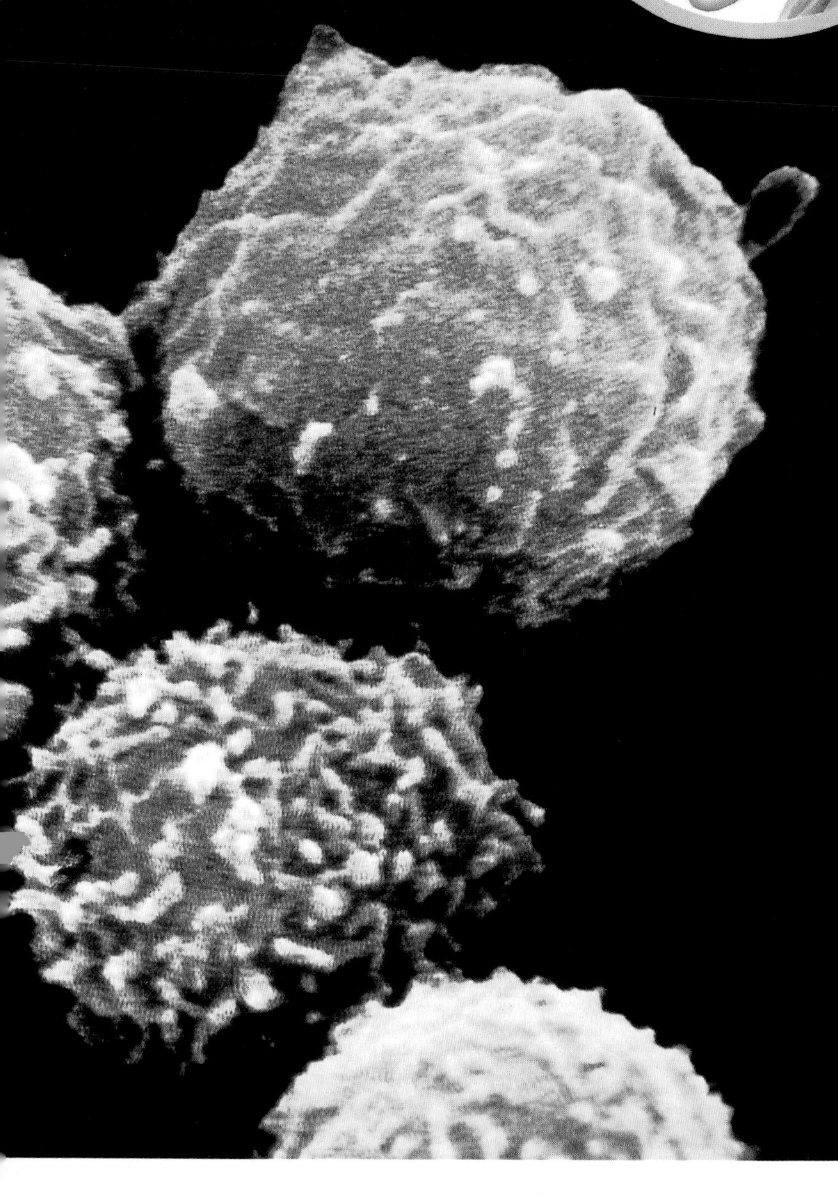

4

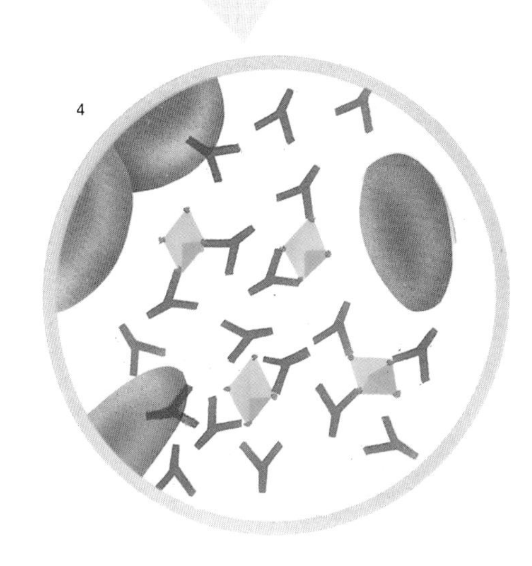

Monoclonal antibodies

Although scientists knew that B cells produced antibodies, at first they could not harness their productive power. The B cells that scientists produced in artificial culture in fermentors (➧ page 83) would survive for a few hours, before ceasing antibody production or dying.

The breakthrough came in 1975 when Georges Köhler (b. 1946) and César Milstein (b. 1927) working at the Medical Research Council Laboratories in Cambridge, UK, developed hybridoma cells. These are fusions of B cells and cancer cells. The cancer cell enables the fusion to multiply and grow in fermentors for long periods, while the B cell part enables it to produce antibodies as it does so. All the antibodies produced by the hybridoma cells are identical, and are the same as those produced by the B cell before it was fused. Thus scientists can specify which antibodies they want by choosing an appropriate B cell for the fusion. Because the antibodies that are produced all come from a single clone of hybridoma cells, they are called monoclonal antibodies.

▲ *Dr César Milstein who, with Georges Köhler, received the Nobel Prize for Medicine for his work developing hybridoma cells. Hybridoma cells produce monoclonal antibodies which have been used to diagnose disease, purify drugs and as an experimental treatment for some forms of cancer.*

Since Köhler and Milstein's discovery, hundreds of ways of using monoclonal antibodies have been found. Medical and veterinary diagnostic tests (➧ page 98) form a particularly important category of products. The binding power of monoclonal antibodies is also useful in extracting other protein products, like interferon, from cultures of genetically engineered organisms. Through genetic engineering, biotechnologists are now attempting to produce improved "second-generation" monoclonal antibodies. Antibodies successfully bind to antigens, but outside the body they can do little else. To remedy this, scientists have spliced together a gene that codes for the antigen-binding arms of the antibody with a gene coding for an enzyme. The hybrid antibody-enzyme molecule still binds to its target antigen and can be detected by tests for enzyme activity.

Doctors hope to be able to use monoclonal antibodies for purposes other than diagnostic tests, in particular, to attach to them toxic drugs that will bind cancer cells. Injected into a cancer patient, the combined molecule would circulate around the body in the blood until it found the tumor and then deliver the drug to its target. This treatment would greatly reduce the very traumatic side-effects of present cancer therapy. The problem is that most monoclonal antibodies come from mouse and rat hybridoma cells and therefore have a "rodent imprint". When injected into a patient, they are recognized as nonhuman foreign matter, triggering an immune response that clears them from the bloodstream before they reach the target.

Ideally, the answer would be to develop human monoclonal antibodies. But to obtain human B cells specific for tumors, scientists would have to inject cancer cells into humans, which is ethically unacceptable. Researchers have tried, through genetic engineering, to reduce the amount of rodent protein in the antibodies. In one experiment, they combined the genes for human constant regions with those for mouse variable regions. In a second case, just the mouse hypervariable regions were used, with the rest of the protein coming from human sources. In both cases, the specificity of the hybrid antibody was determined by mouse genes while the rest of the molecule was human. These antibodies have not been tested clinically.

Another problem is that antibodies are unnecessarily large for the binding function. Accordingly, there are plans to produce much smaller molecules, consisting of the antibody-binding variable regions alone, by splicing together the V genes from light and heavy chains.

Catalytic antibodies

Scientists have asked whether antibodies could be produced that would act like enzymes to catalyze biochemical reactions. The conversion of a substrate by an enzyme can be imagined as a two-step process. Enzymes rearrange the atoms of the substrate so that it ends up as the product. While this is taking place, a transition-state molecule, a hybrid of the substrate and product, occupies the enzyme's catalytic site. Scientists reasoned that antibodies against the transition-state molecule might also convert substrate to product. Unfortunately, this theory cannot be tested, because transition-state molecules are extremely short-lived. However, transition state analogs, long-lived molecules that are transition-state "lookalikes", do stimulate the production of antibodies, which sped up the particular reaction for which they had been produced. But these "abzymes" (antibody-enzymes) were slow compared with true enzymes. Nevertheless, abzymes are highly promising as ways of catalyzing reactions for which no true enzyme or chemical catalyst is available.

▼ *Monoclonal antibodies are often needed in kilogram quantities, so companies like Celltech have developed large fermentation vessels in which to cultivate hybridoma cells. The fermentation vessel must remain germ-free. Any contamination would render all 2,000 liters unusable.*

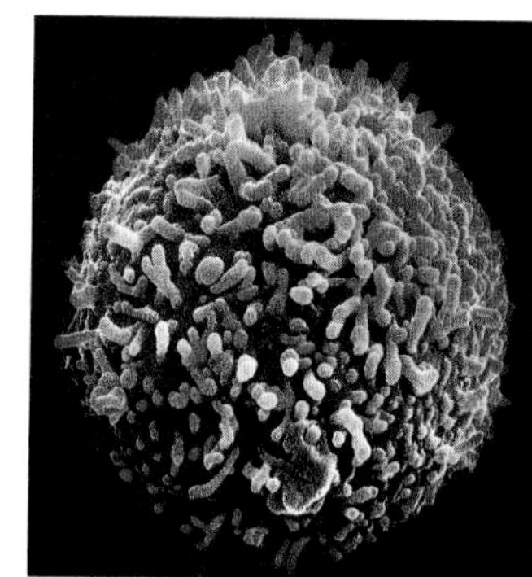

▼ *This hybridoma cell is magnified 2,000 times. Antibody-producing cells are not normally long-lived, but when fused with cancer cells they can live for ever and still continue to produce antibodies. However, because hybridoma cells are derived from cancer cells, those monoclonal antibodies intended for medical uses have to be highly purified and are subjected to a battery of stringent safety tests.*

The total output of manufactured enzymes is tiny compared with that of naturally occurring enzymes

Enzyme production

Biochemists have identified more than 3,000 enzymes, but only a handful of these are produced in large quantities. The most abundant enzyme is ribulose-bis-phosphate carboxylase/oxidase, called RuBisCo for short. This is the enzyme which plants use to capture carbon dioxide from the atmosphere in order to convert it into organic matter. Plants produce many million tonnes of RuBisCo per year. Human efforts are puny in comparison. The total annual output of all the major enzyme-producing firms is around 1,300 tonnes. With the exception of glucose isomerase, all the enzymes produced in large amounts are hydrolases – that is, enzymes that split natural polymers into smaller pieces.

The biggest portion of this total, 500 tonnes, representing 40 percent of worldwide enzyme sales, is accounted for by protein-digesting enzymes (proteases) produced by bacteria and used in "biological" washing powder. Proteases break up protein components of stains such as from blood, egg, chocolate and milk, and help to dissolve soil. The enzymes are so active that only 1.5–2.5 grams need to be added to every 10 kilograms of washing powder. Some proteases can withstand wash temperatures of up to 70°C and resist destruction by bleach and other additives. The cost of adding enzymes works out at about 4 US cents per kilogram of washing powder.

Two enzymes which break down starch, alpha-amylase and glucoamlyase, are also produced in bulk (each around 300 tonnes per annum). Starch extracted from plants is a starting material for many industrial processes. Alpha-amylase is used to control the viscosity of starch solutions to provide a gloss finish to paper or paints. Glucoamylase splits off glucose molecules from starch – an important step in the production of industrial alcohol and the first stage in the manufacture of high fructose corn syrup (HFCS) sweetener. The second stage of HFCS production requires another enzyme, glucose isomerase, to convert the glucose into fructose. One kilogram of glucose isomerase will produce up to 9,000 kilograms of product. Industrially used enzymes are robust and may retain up to half their initial activity after two months of continuous use.

Enzymes are produced by microorganisms growing in fermentors (♦ page 83) which can have a capacity of more than 100,000 liters. The microorganisms, usually strains of Bacillus bacteria or the fungus Aspergillus, grow on cheap nutrients such as molasses, potato starch, bran, soybean meal, fish meal, or corn steep liquor, which are essentially waste products from other industries. In the future, genetic engineering may realize the possibility of growing fermentation organisms on even cheaper materials such as cellulose from straw.

▶ Many governments are concerned about industrial wastes in effluent and gaseous emissions from factories. Accordingly, companies are developing on-site waste clean-up systems. In the Netherlands, where the density of population and dependence on freshwater canals make pollution problems particularly acute, the Gist-brocades company cleans its waste water by trickling it through towers containing huge numbers of bacteria. The bacteria convert waste products into methane which the company burns in its boiler-rooms.

Biotechnology and Agriculture 4

Plants as the primary food source...Developing strains by backcrossing...Plant genetic engineering ...The Ti plasmid vector...Engineering herbicide resistance...Improving quality and output of food... Plant cell regeneration... Somaclonal variation.... Fermentation old and new... Microorganisms in biotechnology and in nature...Animal vaccines ...Genetically engineered animals

▲ The Tassili rock painting in Algeria depicts farming in ancient times. With the advent of agriculture, people started to take control of their own destiny. By domesticating livestock, they turned the strength and skills of animals to human ends.

▼ Modern types of wild barley bear a close resemblance to the remnants of varieties found by archeologists at sites dating back 10,000 years. The varieties of barley used today are descended from wild strains but have much larger grains.

Plants and their genes

Plants are the primary sources of human and animal food. Fueled by light energy from the Sun and assisted by a range of helpful microorganisms, they convert carbon dioxide and nitrogen gases from the atmosphere, and rocks and water from the Earth, into their own tissues. Animals (including humans) are incapable of doing this. We cannot obtain the chemicals we need for growth and maintenance simply by breathing air, eating soil, drinking water, and lying in the sun. We must eat plants or animals who have themselves eaten plants.

Growing plants is the world's oldest, largest and most important biotechnology. Without it, only a small fraction of the human population could survive on Earth. We depend on plants for paper, furniture, fuel, many medicines and food. Most vitally, we rely on plants for the oxygen they release into the atmosphere.

For thousands of years, humans have selected plants to meet their own needs. Oranges are juicier, cereal grains are larger, forests are faster-growing than they were before humans intervened. Modern crops are high-yielding and chosen for their ease of mechanical harvesting. Varieties have been produced to be more resistant to summer drought or winter cold, making growing seasons longer and enabling farmers to sow crops over wider geographical areas.

Plants are carefully interbred so that desirable characteristics from different varieties are combined in a new strain. Agricultural crops are artificial, in that they would not survive well in the wild. Until recently, plant breeders had to depend on the natural mechanisms of reproduction. To improve the disease-resistance of a high grain-yielding commercial strain, pollen from a disease-resistant wheat, possibly from among wild varieties, might be used to fertilize the commercial strain. The resulting plants would have a mixed genetic composition – genes from both parental plants – and would display mixed characteristics, some good, some bad. The undesirable ones could be gradually removed from the new plants by repeatedly backcrossing, that is, by breeding them with the high-yielding parent. Eventually, a high-yielding, disease-resistant plant with no undesirable features would be produced.

Selecting, crossing and backcrossing in this way can take many months, and the entire breeding cycle, many years. As well as being time-consuming, traditional plant breeding is limited in its scope for improving plants. Genetic mixes can only be made between plants that can be interbred. When genetic engineering techniques were developed for plants, however, genes no longer had to be mixed and sorted randomly. The process could be directed. Furthermore, genetic engineers could take the desired genes from any source, not just from reproductively compatible plants. The method is also very much quicker than the traditional way of breeding new plant varieties.

► ***The skills of "nature's genetic engineer" being shown to journalists. The cankers at the base of the sunflower are the result of infection by Agrobacterium tumefaciens, a type of microorganism found in the soil. To cause the canker, the bacterium inserts some of its own DNA into the plant's chromosome. Human genetic engineers have learnt how to use Agrobacterium to ferry foreign genes into plants.***

▲ ***A scientist from Agrigenetics, Inc., one of the leading US plant biotechnology companies. Plant genetic engineers have found out a lot about the genetic vector Agrobacterium tumefaciens. In particular, they have worked out the DNA sequence of part of its Ti plasmid, a small circle of DNA which is responsible for the cankers caused by Agrobacterium. The scientists replace these canker-causing genes (the sequence of one is highlighted in green) with DNA which codes for functions useful to farmers. A root-infecting relation of Agrobacterium tumefaciens has also been used in genetic engineering.***

Although plants are far more complex than microorganisms, the principles of genetic engineering are much the same in both. The gene (or genes) isolated from the donor organism is inserted in a carrier molecule of DNA – the vector. This vector is then introduced into the plant cell where it gives rise to the protein coded for by the inserted gene. The difference lies in the origin of the vector and how it finds its way into the plant cell.

The most often used carrier of genes in plant genetic engineering is the Ti plasmid, a bacterial plasmid that can also function in plant cells. It comes from a remarkable bacterium called *Agrobacterium tumefaciens* which lives in the soil. This is a plant parasite which infects scratches or cuts in plants, causing a cankerous growth called a crown gall. It does so by invading plant cells and releasing its Ti plasmid. Part of the plasmid DNA inserts itself into the chromosome of the plant where it is treated like a normal plant gene. The inserted DNA converts the invaded plant into a factory producing food for *Agrobacterium*. In this way the infection is sustained. Scientists have called *Agrobacterium tumefaciens* "nature's own genetic engineer".

Human genetic engineers have adapted the *Agrobacterium* Ti plasmid to transfer foreign genes into plants. Using test-tube techniques, they remove the genes causing the canker on the plant but retain those that code for insertion of the DNA into the plant chromosome. Then they splice new genes into the "disarmed" plasmid, before putting it back into *Agrobacterium*, which is then used to infect plants. Because this new version does not cause crown galls, however, genetic engineers cannot easily tell which plants have been infected. Therefore, one of the first genes to be transferred into plants is coded for resistance to an antibiotic called kanamycin. Watering the plants with a solution of kanamycin ensures that only successfully engineered plants, those with a resistance gene, will grow. Similar methods can produce plants with drought tolerance, immunity to insect pests and resistance to herbicides.

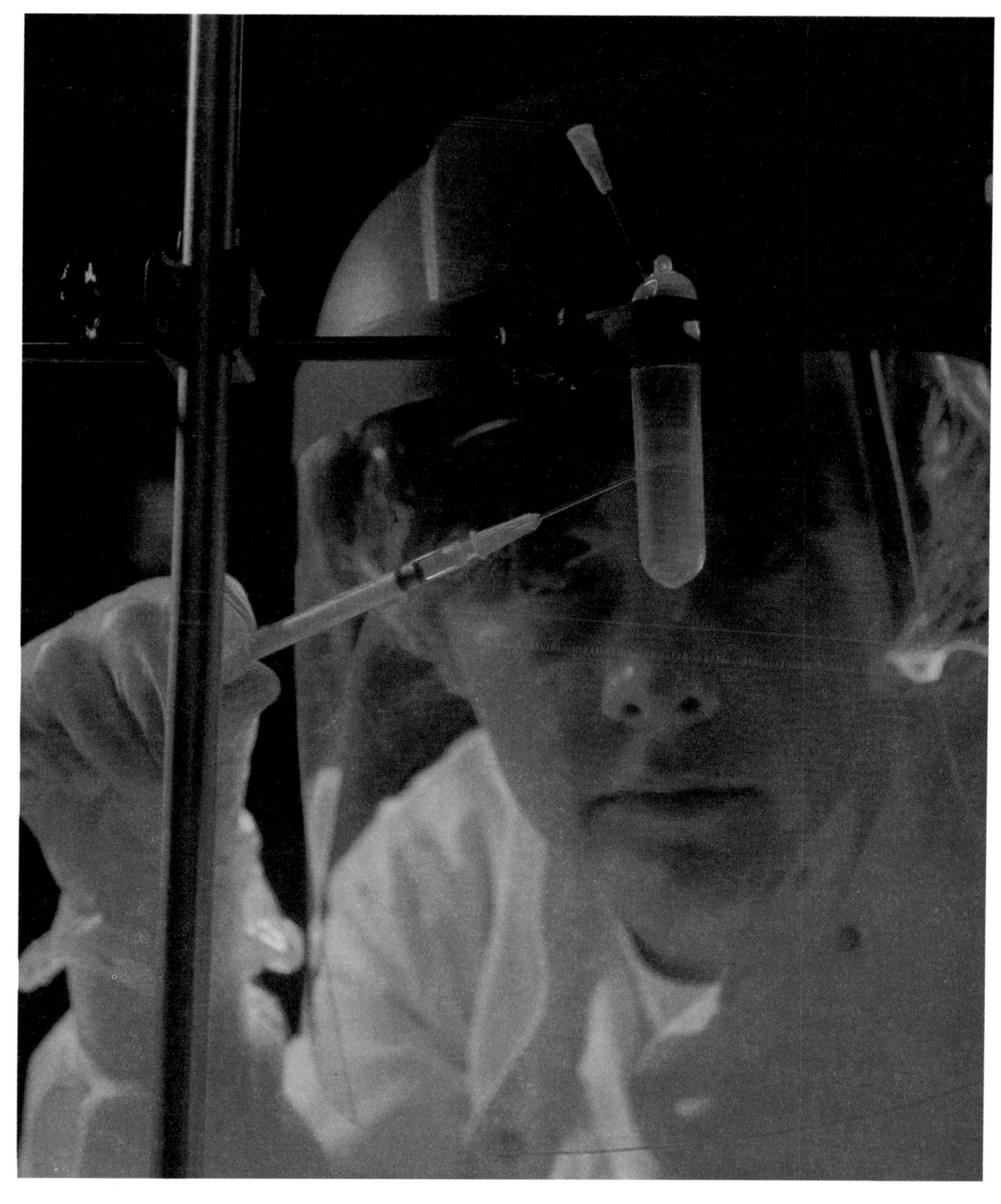

◀ *A technician carefully extracts a genetically engineered plasmid through the wall of a plastic test-tube. Very high speed centrifuges spin the tubes. These separate the DNA of the bacterial chromosome (upper band) from the slightly less dense plasmid DNA. A plasmid purified in this way from one organism can enter a different microorganism, taking with it any inserted genes.*

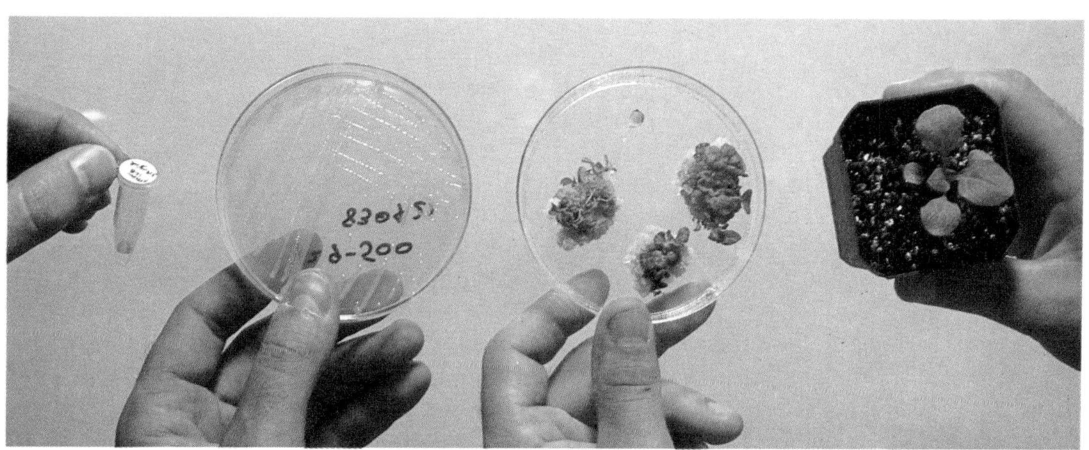

◀ *Putting genes in plants. The small tube contains DNA coding for resistance to the antibiotic, kanamycin. Cultures of Agrobacterium (second left) ferry the DNA into tobacco cells. Tiny genetically engineered plantlets develop, which after a few weeks can be transferred to soil.*

Under natural circumstances, *Agrobacterium tumefaciens* only infects broad-leaved plants such as tobacco, petunia, tomato and rape. The crops most important to world agriculture – wheat, rice and maize – are not susceptible. There are other ways of introducing foreign DNA into these crops. In most of them, the plant is first stripped of all its protective barriers. Leaves, for instance, are macerated into very small fragments in a food blender. Treating the resultant pulp with enzymes then removes the walls of individual cells. In the wall-less cells, or protoplasts, only a fragile barrier to the entry of DNA, the cell membrane, remains.

The membrane can be breached in several ways. One is to mix two protoplasts from different plants and induce them to fuse together. Such a method produced the pomato, a potato-tomato hybrid, which contained DNA from both parents. The mix of genetic material was random and the pomato was sterile. Usually, DNA manipulation is more directed and more selective. A method called electroporation has been particularly successful. A mixture of protoplasts and purified DNA (from any source) is placed in a pulsating electric field. The field causes holes or pores in the protoplast membrane to open up temporarily through which the DNA can pass. Chemical treatments can also be used to breach the membrane. In both cases, DNA must enter the plant nucleus and become integrated into the plant chromosome. Another method uses extremely fine needles to inject DNA into plant protoplasts. With even more intricate manipulation, DNA can be inserted directly into the cell nucleus.

The next problem is to regenerate whole plants from the fragile protoplasts, many of which are extensively damaged by the relatively harsh chemical, physical or electrical treatments. Some protoplasts can regenerate easily, others not at all. The main difficulty is in the first stage of regeneration, the regrowth of the cell wall, which scientists do not yet fully understand.

Herbicide-resistant crops were among the first targets for genetic engineers. Herbicides reduce the weeds and ensure that as much of the nutrients as possible reaches the crop. Glyphosate is a herbicide used widely by both farmers and gardeners under such tradenames as Round-up and Tumbleweed. It works by blocking an enzyme that is in all plants and is essential for their growth. Glyphosate, therefore, like most herbicides, kills all plants, weeds and crops alike, so it must be applied before the crop emerges. But this is less effective for weed control than spraying during the growing season would be (weeds also grow fastest during the growing season).

For this reason, biotechnologists have developed plants that are reistant to glyphosate. They used as a source of genes a glyphosate-resistant strain of the gut bacterium, *E. coli*. They isolated the genes coding for the enzymes responsible for the resistance and, using *Agrobacterium*, transferred them to plants such as tobacco, petunia and rape. The genetically engineered plants could grow when sprayed with herbicide. Herbicide resistance can also be conferred on potatoes, tomatoes and tobacco by inserting appropriate genes.

Herbicide resistance will allow crops to grow unhindered by competing weeds and yields per hectare are likely to increase. However, some environmentalists are concerned that an increase in the use of herbicides may bring with it an increase in chemical pollution. The "agrochemical" companies producing the herbicides disagree. The ability to confer resistance on plants, they say, will allow farmers to choose herbicides that are more environmentally benign.

▶ Winter and spring wheat maturing on a typical prairie farm. The world's most important crops, wheat, rice and maize, are cereals. Cereals are not naturally susceptible to infection by Agrobacterium tumefaciens and so have proved difficult targets for genetic engineering. This has spurred scientists to develop other, physical, chemical and electrical techniques for getting new DNA into plants.

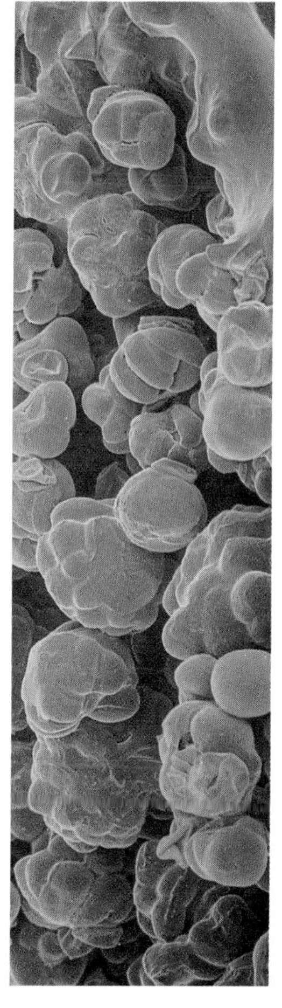

◀ *Grains of barley pollen in artificial culture. Under special nutrition, pollen grains mature into plants with only one set of chromosomes (instead of two). However, the chromosomes can double up, to provide true-breeding plants rapidly.*

▼ *Blasting DNA into plants. A microscopic tungsten pellet (blue dot) which bears a cargo of DNA has been fired at short range and high velocity from an experimental cannon into onion cells. The white area surrounding the pellet is the zone of activity of an enzyme which is coded for by the DNA. The existence of the zone shows that the cells and the DNA on the pellet are still functioning.*

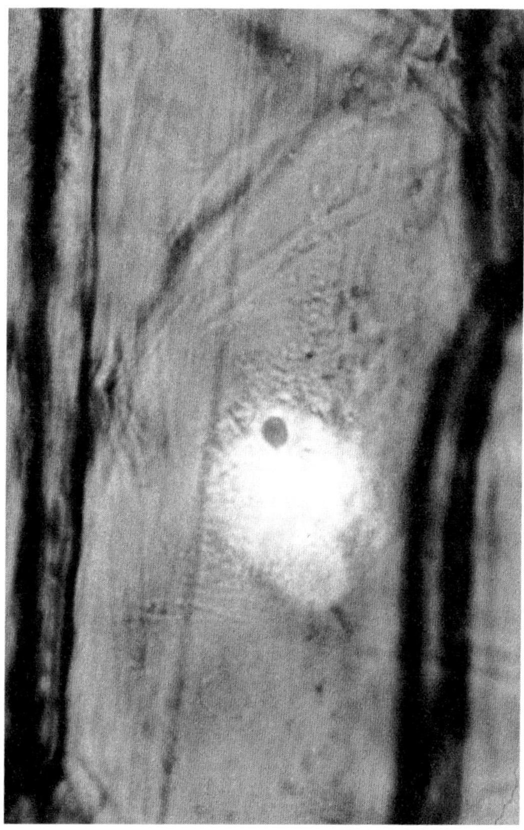

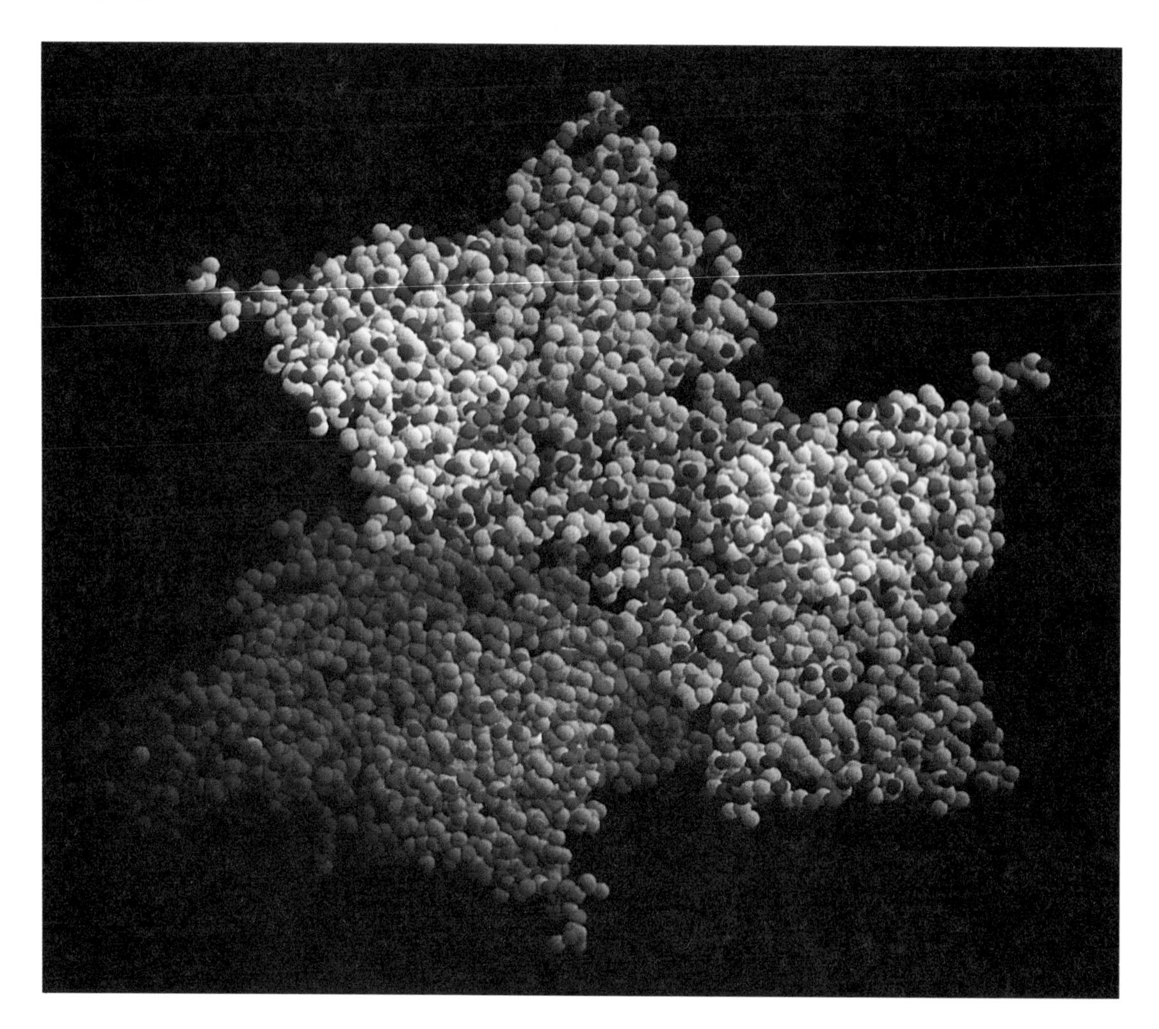

► ***The protein coat of the virus that causes bushy stunt disease in tomatoes. Scientists have discovered that by inserting genes for protein coats of viruses into the DNA of tomato plants, these, in effect, vaccinate the plants against the disease. Although successful, no one fully understands how the protection works.***

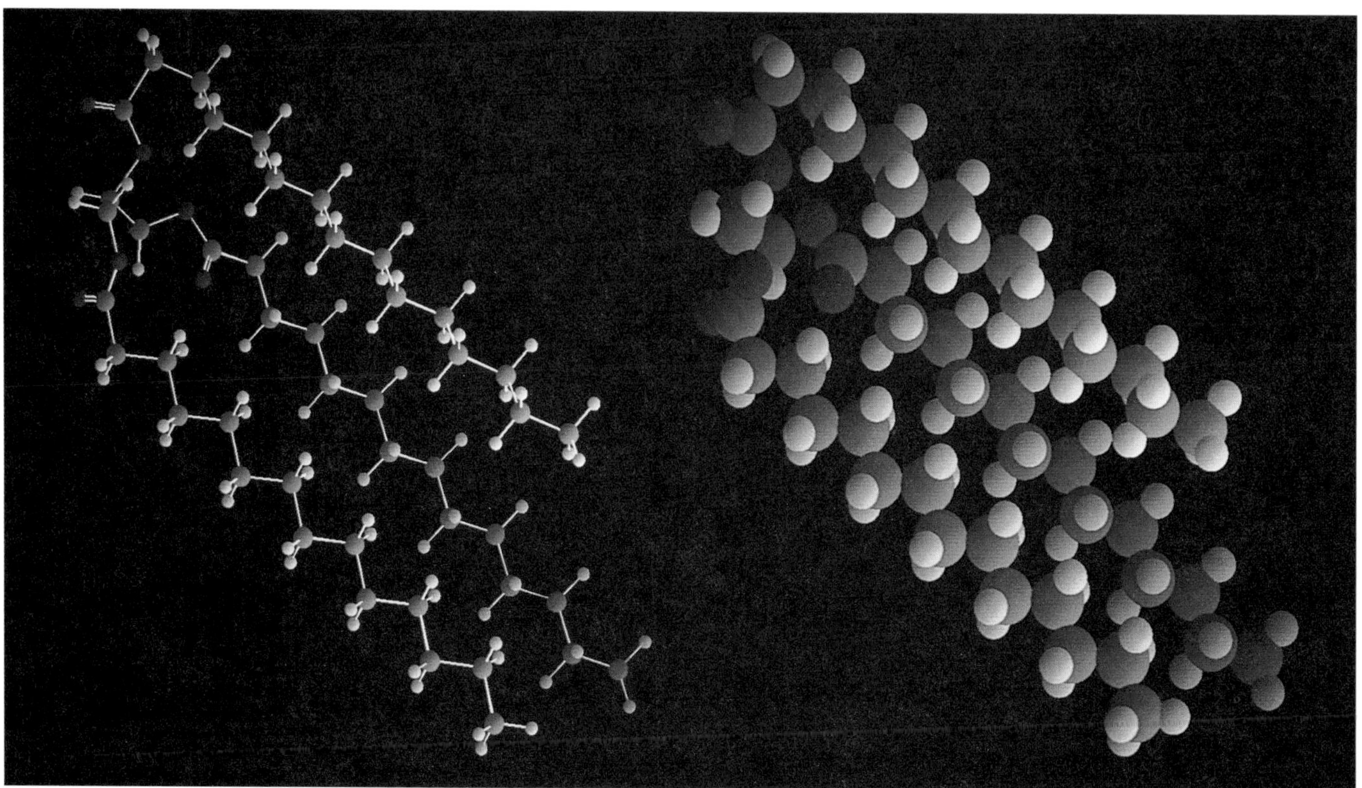

Genetic engineering may remove the need for plants to be sprayed with insecticides. The toxic protein from the bacterium *Bacillus thuringensis (B.t.)*. kills larvae of butterflies and moths but has no effect on animals or plants. In 1987, scientists in Belgium and the USA took the gene coding for the *B.t.* toxin from the bacterium and inserted it into tobacco plants via the *Agrobacterium tumefaciens* Ti plasmid. The plants produced the toxin and killed predatory caterpillars but otherwise grew normally.

Viruses are another source of crop damage and one for which no chemical treatments or cures are available. Genetic engineering may again provide a solution. Tobacco mosaic virus, which produces a mottled appearance in tobacco leaves, consists of a spiral of RNA surrounded by a protein coat. If the gene for the coat protein is inserted into the DNA of cells from the tobacco plant, the plants which result are immune to infection from the virus. The same result is seen in tomato plants. Scientists do not understand why this occurs but it could be a very useful way of preventing crop diseases.

◀ *The laboratory farm. Research on genetically engineered plants is conducted inside special growth rooms where the illumination, humidity, temperature and atmospheric conditions can be precisely controlled. It is only after tests under these conditions that plants will be grown in greenhouses or outside in fields.*

◀ *All oils and fats are triglycerides, molecules in which three chains of fatty acid are linked to glycerol. In unsaturated fats, such as butter and lard, the fatty acids are straight. Neighboring fat molecules tend to line up in rows making the fat solid. In plant fats, which are unsaturated, or polyunsaturated, the fatty acids are bent at one or more positions and the fats are usually liquids (oils).*

Genetic engineering can also improve the nutritional quality of plants. Protein deficiency is the most common form of malnutrition, particularly in countries where one crop provides 80–90 percent of the population's protein intake. Human beings can only make 10 of the 20 amino acid building blocks of proteins. The rest, referred to as the essential amino acids, must be provided in the diet. But some crops have only very low levels of essential amino acids. Generally, plant protein is stored in the seeds (beans, grains and nuts). Roots, stems and leaves are often discarded. By targeting proteins rich in essential amino acids to the seeds, the nutritional value of plants can be increased.

For instance, the proteins of many legumes (plants of the pea and bean family) that are the staple diet in parts of South America and Africa, are deficient in essential amino acids containing sulfur. This imbalance could be corrected if the genes coding for the seed proteins of the Brazil nut (which contains up to 25 percent of these amino acids) were introduced into legumes. The potato, another staple crop, has been engineered with a synthetic gene coding for high levels of several essential amino acids. The gene was produced on a DNA synthesizer (◀ page 32) and inserted into the plant using the *Agrobacterium* Ti plasmid. Calculations showed that the new protein doubles the potato's nutritional value.

Many crops are grown for their oil or fat content. Palm and coconut provide the raw materials for soap and detergents. Olive, sunflower, peanut, soya and canola oils are edible. Rapeseed, castor and jojoba oils are used industrially as lubricants. All oils are triglycerides, in which three long, fatty molecules (fatty acids) are joined, like the prongs of a fork, onto a molecule of glycerol. The type of fatty acids and their arrangement in an oil depend on the enzymes present in the plant cells. Since genetic engineering will enable gene exchange, new plants capable of producing any desired oil could be bred. For instance, if scientists understood how the cacao plant controls the production of cocoa butter, they could produce a substitute by transferring genes from the cocoa plant into soybeans. The genes would code for enzymes involved in fat production. Soya can be grown very efficiently in temperate climates and would be a much cheaper raw material for chocolate than the cacao tree. Similarly, if palm and coconut oils used to make soaps could be produced in rapeseed, it would greatly reduce the need for the tropical crops (▶ page 120).

Cloning from single cells offers a way of rapidly increasing plant production

Plant cells: regeneration, variation and culture

Gardeners frequently clone plants. A branch or a stem cut from a shrub or flower and treated with a rooting mixture will often give rise to a whole new plant. Similarly, "seed" potatoes when planted will yield next year's crop. They are not seeds, merely specialized parts of the root of the potato plant. This sort of cloning may not seem very remarkable, but in humans it would be the equivalent of a whole person growing from an amputated foot. "Totipotent" is how scientists describe the ability of plant tissues to develop into whole organisms.

Biotechnologists have taken the art of plant cloning one step further. They have have reduced plants to their smallest possible size – single cells – and from these cells they have been able to regenerate whole plants. In principle the process is simple. It is only complicated in practice by the need to conduct all operations under germ-free conditions. A piece of plant tissue (about 0.5cm square) is sterilized chemically to kill microbes on its surface and then treated with enzymes to remove structural components like cellulose. This produces a solution containing thousands of individual plant cells. The cells can grow in sterilized liquids in glass flasks (as long as appropriate nutrients are supplied) or in Petri dishes. Hundreds of thousands of cells, each capable of developing into a whole plant, can be grown in one vessel.

Plants are regenerated by growing cells on a gel medium containing certain plant hormones as well as nutrients. The cells are colorless and grow initially as a random mass called a callus. After a few days or weeks, tiny green buds, leaves or roots appear. This process is called somatic embryogenesis – the founding of embryos from the somatic (body) tissue. The number of embryos appearing increases tenfold when small electric currents are passed through the tissue, using a

▼ Plant biotechnologists have produced new varieties of ornamental plants using cell and tissue culture methods. After plant breeders first develop a new type of flower, suppliers usually need a long time to produce enough seed or plants to sell. Now some plant companies have managed to reduce this delay by macerating the new plants into small pieces of tissue and regenerating whole plants from the pieces. Strict hygienic standards ensure germ-free plants.

► To ensure that "super plant" varieties find their way to farmers, a company in California has developed synthetic seeds, capsules (bottom right) which contain very small pieces of plant tissue. They could also contain bacteria (the contents of the test-tubes) which help the plants to grow. Millions of synthetic seeds can be rapidly produced from a small number of genetically improved plants such as the "super celery" shown. Seeds of crops such as wheat, rice and maize could also be produced in this way.

9-volt battery connected through a resistor to two wires. After sterilization, one of these is inserted into the callus, the other into the jelly. Scientists believe the electrical current endows the formless callus with a sense of direction.

The miniature plantlets can then be transferred to individual sterilized containers where their development continues. After a few weeks or months, they attain sufficient size and strength to be planted in compost to harden them off before they are transferred to the soil. An interesting alternative to this gradual nurturing is to coat the plantlets at the embryo stage with nutrient jelly and a protective outer film. These artificial seeds can be planted directly in soil using adapted agricultural sowing equipment. Cloning from single cells provides a very quick way of multiplying numbers of plants. When new plant varieties are developed, large quantities can be made available to farmers much sooner than was possible in the past.

▼ ***Bananas have three sets of chromosomes and this makes them sterile – they cannot produce fertile seeds. New banana plants, therefore, have to be produced by cloning. The old way of cloning was to plant out the suckers which appear periodically from existing plants. Now, instead, small pieces of tissue from the plants are grown in test-tubes on an artificial jelly containing nutrients and hormones (inset). When the plantlets have rooted, they are transferred to pots and hardened off in hothouses before being sent to plantations.***

Genetic engineering is not the only way of improving plants by the use of single plant cells. When experimenters first grew plants from individual cells, they expected that the plants would all be identical – identical to each other and to the parent plant from which the cells had been taken – clones, in other words. To their surprise, the plants displayed a high degree of genetic variation. Scientists now think either that the individual cells in a plant are different or that differences are induced by the process of culturing single cells. If plants grown from single cells are themselves broken down into single cells, however, these cells do give rise to identical, or cloned, plants. Cloning by this method is called somoclonal variation. It is not a form of genetic engineering since the genetic changes are due to manipulation of cells rather than direct manipulation of DNA.

Using somoclonal variation, new varieties of plants can be developed within a growing season, much more rapidly than by either conventional breeding or genetic engineering. An American company, DNAP, has already produced a new range of healthfood snacks based on crispier carrots, crunchier celery and low-salt popcorn. A British company, Unilever, has used somoclonal selection to develop a series of new varieties of oil palm, resulting in bigger oil yields and a better balance of oil types in the crop. Through the use of cloning, they were also able to produce uniform trees that could be harvested mechanically rather than in the traditional manner using sharp knives tied to long poles.

Plant cells can be grown in fermentation vats as if they were microorganisms. Under these conditions, their growth is slow and the cells are often fragile but will produce a wide range of highly valuable compounds. The Japanese Salt and Tobacco Monopoly has the world's largest plant cells culture vessel, with an internal volume of 20,000 liters, the size of a small room. The most successful venture has been the production, by another Japanese company, of a purple dye called shikonin from cells of *Lithospermum erythrorhizon*, a plant that was almost extinct in Japan. The dye is normally extracted from roots of plants which are from five to seven years old. It can be obtained from cultivated cells after 20 days or so. It has been used extensively in Japan as a colorant in lipstick and nail varnish, and as a component of many other skin-care products. It also has medicinal value as a treatment for burns and hemorrhoids.

Plant cells are also a potentially good source of flavors, colors and fragrances such as jasmine, mint oil and betanin (the red dye from beetroot). Herbs and spices might be other candidate targets for biotechnology. However, many of the active ingredients in the plants are very simple and can be produced more cheaply by chemical methods. On the other hand, some fragrances and flavors are extremely complex mixtures of hundreds of separate compounds which would be difficult to reproduce in culture.

One of the attractions of plant cell culture is that substances produced in this way may be classified as "natural products", making their use in foodstuffs more acceptable. It is also a more flexible and predictable way of making plant products than agriculture. Many of the most useful medicinal crops grow only under particular climatic conditions. For many countries, plant cell culture will reduce reliance on imports, free supplies from political influence by ensuring that vital drugs can be produced domestically and eliminate the effects of disease and climate. In addition, products may be more consistent and, as the technology develops, become cheaper.

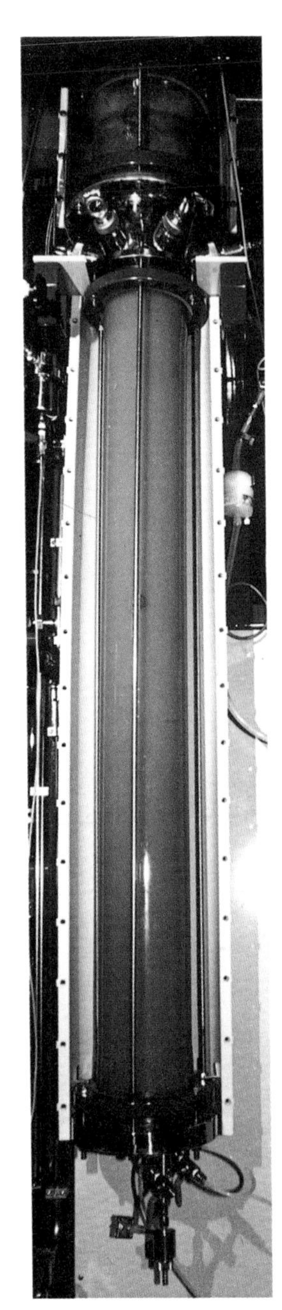

▶ The "souper" tomato. Using somaclonal variation, scientists produced a type of tomato that was much more solid than normal. Although not ideal for salads, this variety was useful to soup and ketchup manufacturers who are primarily interested in a tomato's solids, not its water content.

◀ Periwinkle cells grown under controlled conditions in an airlift bioreactor. Nutrients for the cells enter the fermentor through inlets at the top of the 2-meter glass column. Air bubbles enter at the bottom and, as they rise to the top, they create a flow which gently mixes the culture.

◀ *Yellow gentian growing in Switzerland. Many plants, like gentian, have long been valued for their medicinal qualities. Chemists have used organic synthesis to manufacture drugs based on plant compounds. Now those drugs and others can be produced in cultures of plant cells.*

◀ *Many drugs have been produced by cultured plant cells. The pink periwinkle produces vinblastine and vincristine, compounds used in the treatment of leukemia. The plants grow slowly, making it difficult to obtain sufficient quantities for the drugs. Cultured cells, on the other hand, double their numbers every day and will still produce the drugs. Antispasm drugs, sedatives and heart stimulants have all been produced in plant cells. Foxglove cells can convert a low-cost chemical, digitoxin, into the highly valued heart drug digoxin.*

Fermentation, an ancient method of preserving food, has been developed by biotechnologists to produce drugs and other valuable substances

► **Brewing, one of the earliest forms of fermentation. Biotechnologists have updated traditional fermentation methods. Today, fermentation includes any process in which microorganisms (usually) are grown in large amounts to produce any kind of product.**

Traditional fermentations

The earliest fermentations were uncontrolled, dirty and, to begin with, wholly accidental. However, they served the important purpose of preserving the food produced through agriculture. Fermentation extended the productivity of farming, with farmers learning not only how to grow food but also how to keep it for themselves. Food for humans is also food for microorganisms and, left untreated, most agricultural products quickly spoil. The growth of harmful microorganisms will make them either unpalatable or poisonous.

Fortunately, other microorganisms can be marshalled through fermentation to help preserve food. The first fermentation processes, brewing, bread making and cheese making, encouraged the growth of beneficial organisms at the expense of pathogenic organisms. As the harmless organisms thrived, they produced conditions (like high acidity or high-alcohol concentration) in which other organisms could not grow. Furthermore, the growth of the harmless organisms altered the nature of the food, creating new tastes and textures and often entirely new foods.

Modern fermentations

Fermentations developed in the modern era to exploit the biochemical power of microorganisms for the production of particular compounds, such as amino acids or antibiotics, in an electronically controlled, clinically clean, highly defined environment – the fermentor (also called the reactor or bioreactor). Most industrial fermentors are stainless steel or glass cylindrical tanks, commonly from 10 to 10,000 liters in capacity. When in use, they contain liquid in which a single type of organism grows. There may be 10,000 million identical cells in every liter of the liquid, each one of them producing the same useful products. To produce penicillin, biotechnologists use a fungus called *Penicillium chrysogenum*. For amino acids, they might choose a bacterium of the *Corynebacterium* genus, while for alcohol the yeast *Saccharomyces cerevisiae* is most often used. Biotechnologists have also learned how to grow plant cells and animal cells, including those from humans, in fermentors.

Biotechnologists start fermentation processes by adding a few million cells to the sterilized liquid in the fermentor. The organisms need nutrients (food). The fermentation liquid contains most of these, including glucose as a source of carbon, and ammonia for nitrogen, as well as minerals, phosphorus and vitamins. Plant and animals cells are often particular in their nutrient requirements. Blood serum is frequently added for animal cells, hormones for plant cells.

Oxygen, one of the most important nutrients, is usually fed in as air which is bubbled through the fermentation liquid. Some of it passes from the bubbles into the liquid and is absorbed by the growing culture. One of the most difficult problems in fermentation is supplying enough oxygen to the culture. One way of increasing the amount of oxygen reaching the culture is to mix the fermentor contents. In the commonest fermentor, the stirred tank reactor (STR), the mixing action is similar to that of a food processor. Paddle blades or turbines mounted on a central shaft are driven round at speeds of hundreds of revolutions per minute by an external motor. The bubbles are trapped in the swirling liquid streams, spend longer in the fermentor and so transfer more oxygen to the liquid.

As in a food processor, the violent mixing can "shred" the contents of the STR. The rotating blades cause very strong currents which disrupt fragile organisms. Most fermentation microorganisms are robust enough to withstand this treatment but usually plant and animal cells are not. Therefore, biotechnologists treat these more gently. One way of protecting them is to grow them inside or on the surface of small beads, a process called cell immobilization. Another is to mix the culture less violently, using slow-turning turbines.

Most industrial fermentations are controlled by computers. The flow of information from the sensors in the reaction vessel is continuously recorded by the computer and any necessary changes to the fermentation conditions will be implemented according to programmed instructions. If something starts to go wrong the computer will alert the fermentation technician, who must decide whether the fault can be corrected or whether fermentation must stop.

When genetically engineered organisms are being grown, stringent precautions are taken to prevent contamination. The air and the waste products that leave the fermentation area must be sterilized. The doors are replaced with airlocks. Technicians may have to wear allover suits, which are discarded in the airlock and subsequently incinerated. Access to the facility is limited to specialized personnel.

◀ An experimental vessel used by biotechnologists to determine the optimal fermentation conditions before they try to grow organisms in bulk. All aspects of the fermentation are carefully controlled. The metal cylinders on top of the vessels house motors to drive the stirrer paddles which control how the culture is mixed. Each culture is kept at a constant temperature, using a sensor and heating element. Another sensor actuates pumps to add acid or alkinine in order to maintain constant pH. The flows of gases in and out of the vessels, and their compositions, are monitored, and the volume of the gas being used by the microorganisms is calculated. These data are relayed to a computer. The vessels and the pipework of large fermentors are usually made of stainless steel and they operate in the same way as experimental reactors.

▼ The commonest cause of fermentation failure is contamination by unwanted microorganisms. Every precaution is taken to exclude them. All equipment is thoroughly sterilized before use and all lids and joints are tightly sealed. The main sources of contamination are human beings and their dust. Some modern fermentation facilities are cleaner than operating theaters in hospitals and the technicians who work in them take more stringent precautions than surgeons.

A better understanding of the part played by microorganisms is helping biotechnologists to improve agriculture

The role of microorganisms

Microorganisms are the invisible participants in agriculture. Some become manifest in diseases of plants and animals but most remain undetected. However, their absence would be noticed. They are responsible for providing plants with nitrogen and mineral nutrients, they help animals digest their food and they keep diseases and pests at bay. A better understanding of the part played by microorganisms is helping biotechnologists to improve agriculture.

Microbial inoculants are naturally occurring organisms found in the soil, on plants or in insect pests, that have been isolated, grown in bulk in fermentors and purified. They are used in the treatment of soils and crops as an alternative to chemical fertilizers and insecticides. *Rhizobia* are bacteria that grow in the soil close to the roots of legumes. They invade the hair on the roots of the plant, forming a pink nodule. The *Rhizobium* synthesizes a set of enzymes to convert nitrogen gas from the air into ammonia (a process called nitrogen fixing) while the plant provides the bacterium with carbon compounds and other nutrients. In general, each legume selects its own particular strain of *Rhizobium* from a mixed population in the soil. On a world scale, *Rhizobia* fix as much nitrogen as farmers add to the land as fertilizer – around 100 million tonnes per year.

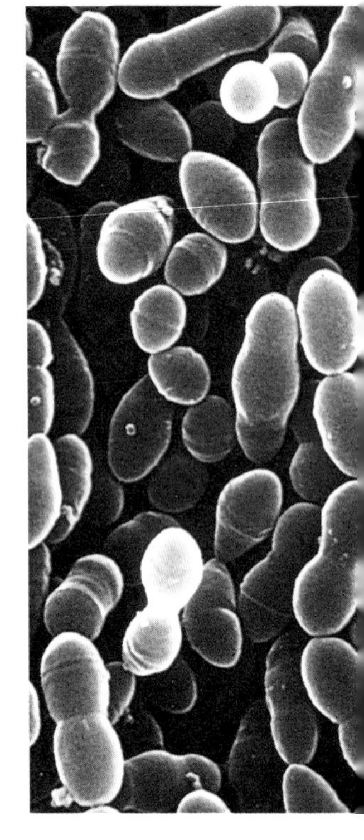

▶ The microorganism Lactobacillus bulgaricus produces lactic acid which is used as an acidulant in many foods. The lactobacillus group of bacteria also help to produce the flavor and texture of yogurt and cheeses. Many types of microorganism contribute to the production and preservation of food.

With a greater understanding of the *Rhizobium* legume symbiosis, biotechnologists hope to improve its efficiency. There are three main targets. Firstly, the nitrogen-fixing reaction itself could be improved. Secondly, the legume could be altered to select the best nitrogen-fixing *Rhizobium*. Thirdly, the process of bacterial–plant recognition and nodulation could be made more efficient. Some scientists believe that it may eventually be possible to induce symbiosis between *Rhizobia* and non-legume plants such as the cereals. If so, it would have an enormous impact on the efficiency of arable farming. One of the difficulties with non-legume plants is that they do not have root hairs suitable for promoting invasion by *Rhizobia*. To overcome this problem, scientists have developed *Rhizobium* strains that can bypass the root hair and invade the root directly. Another solution is to transfer the genes coding for nodulation from *Rhizobia* into *Agrobacterium*, which infects a much wider range of crops, but not the cereals. Ultimately, it may be possible to do without nitrogen-fixing microorganisms by developing plants with an in-built nitrogen fixation based on the systems from *Rhizobium*.

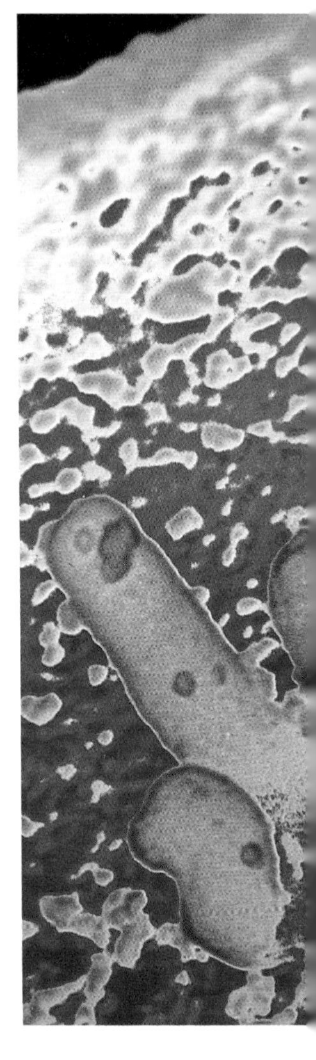

▶ A symbiotic relationship at work. Rhizobia bacteria are attached to hairs on the roots of the pea plant. Later, the bacteria will invade the root hair and help the plant to obtain its nitrogen from the atmosphere. Improving this symbiosis and extending it to a range of crops could enormously benefit those farmers who cannot afford nitrogen-containing fertilizers. Many developing countries are using Rhizobia to improve agricultural productivity.

Another group of microorganisms found near the roots of plants are the mycorrhizal fungi. These grow in the soil as long, fine filaments and help plants to obtain phosphorus, another major component of chemical fertilizers. The fungi act as extensions of the plant's roots, channeling phosphorus (and zinc and copper) from nutrient-rich zones to the plant. Trees inoculated with mycorrhiza grow faster initially and many more survive the trauma of being uprooted.

As well as ensuring that plants are properly nourished, microorganisms can also help to keep them clear of disease by acting as biological-control agents. Cats and dogs are, in a sense, biological-control agents – cats eliminate rodent pests and dogs discourage invasions by burglars. Viruses, bacteria and fungi are commonly used to control insect pests. Intensive forestry has promoted the spread of tree pests. The pine sawfly threatens thousands of hectares of European forest. Fortunately, the fly is susceptible to a virus known as NPV (nuclear polyhedrosis virus), which has been used in biological control. In the past, NPV was grown in insects, but biotechnologists

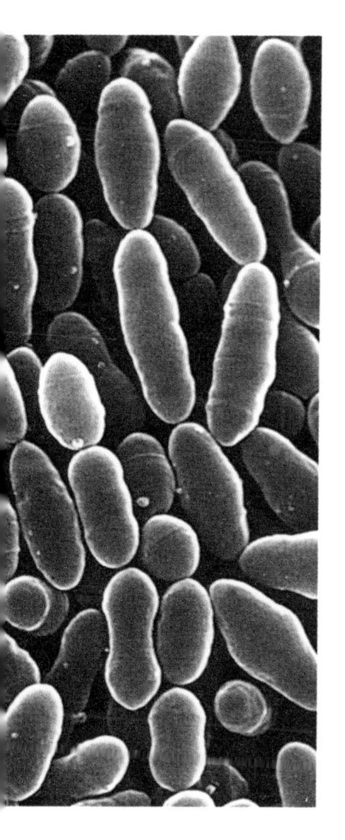

have now developed methods for culturing insect cells in test-tubes and fermentors. These cells provide a much cheaper method of producing the virus. NPV is also being improved genetically. The virus is not selective, infecting beneficial insects, such as those that prevent tree damage by eating other pests, as well as insects that are harmful to the trees. By manipulating its genes, scientists hope to be able to focus its infectivity. In doing so, they may also learn how to adapt NPV when the pine sawfly starts to become resistant to infection.

Damage to plants by frost can also be countered using microorganisms. A bacterium called *Pseudomonas syringae*, which lives in the leaves of plants such as strawberries, makes a protein that produces a nucleus for the formation of ice crystals. Scientists at the University of California have deleted the gene for the nucleating protein to produce an "Ice-minus" bacterium (➧ page 112). When the new strain is sprayed on plants, it will grow on the leaves in place of the normal variety. Small-scale field trials with the new variety have shown that temperatures need to fall much lower than for the normal variety before frost damage ensues. Although ice formation on plants is undesirable, "ice-plus" and "super-ice" strains of *Pseudomonas syringae* have found other applications. Sold under the brand-name Snowmax, they are used to improve the quality and quantity of snow generated mechanically for ski slopes. They could also improve icecream.

▼ How "Ice-minus" bacteria protect plants from frost damage. The beakers contain supercooled water – water below freezing point but still liquid. Normal ("ice-plus") bacteria on the untreated leaf (left) produce a protein which induces the water to crystallize, forming ice. Treating the leaves with "ice-minus" bacteria prevents ice formation. There has been concern that "ice-minus" bacteria might prevent ice crystals in clouds, thereby changing weather patterns.

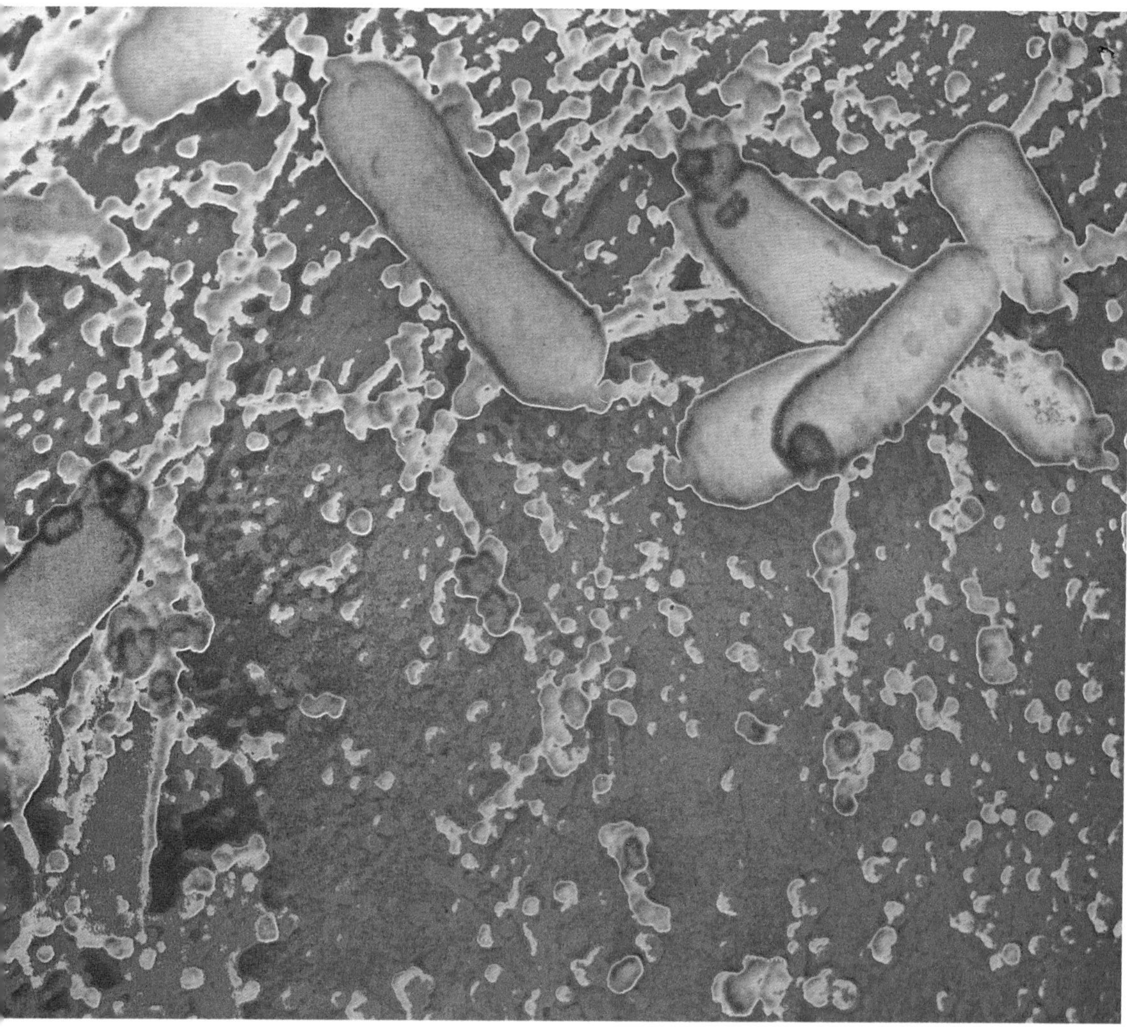

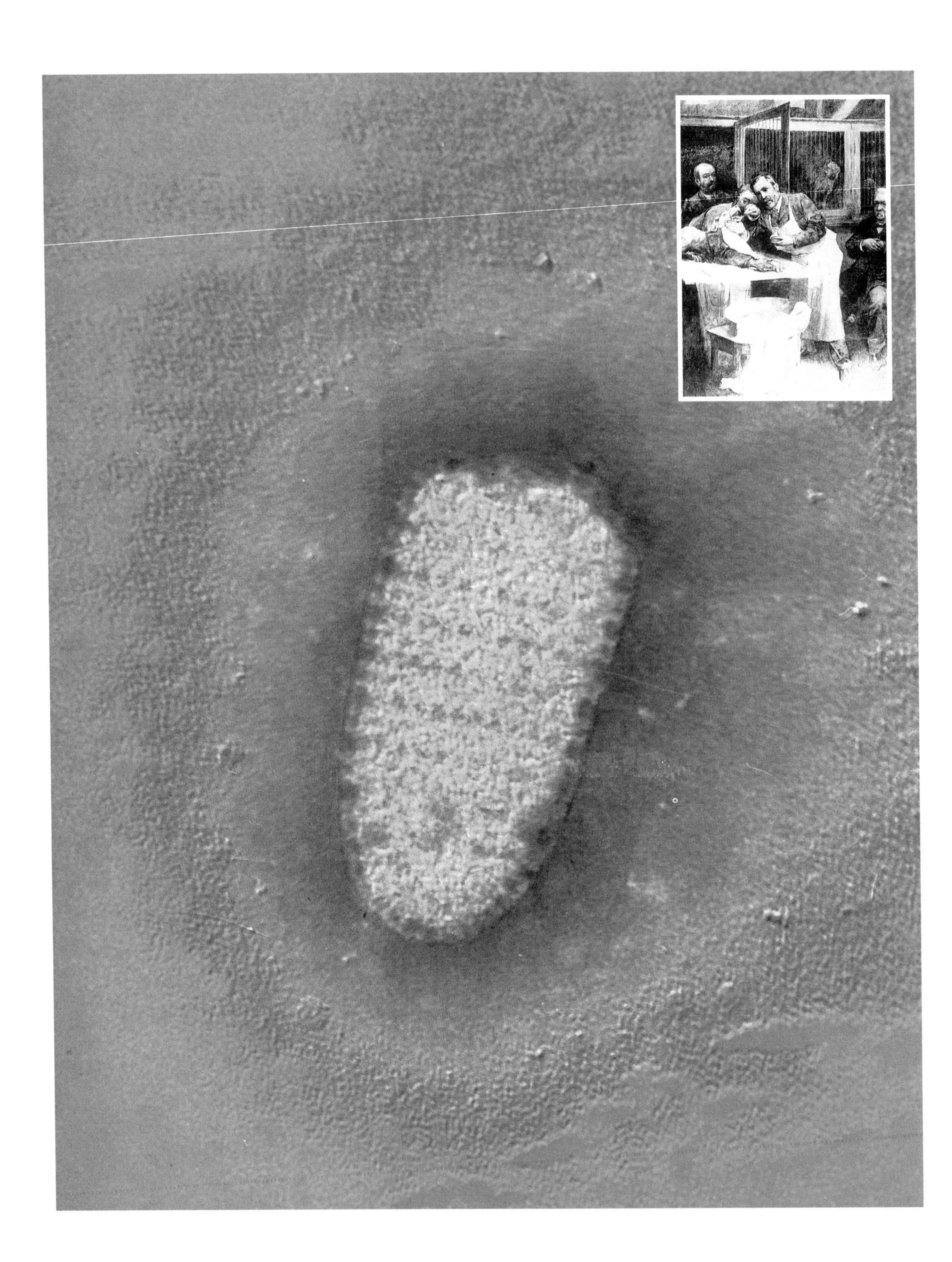

◀▲ (Inset) Louis Pasteur (1822–1895) who produced the first vaccines for rabies, takes a saliva sample from a dog. Today the disease is still prevalent. Rabies is caused by a bullet-shaped virus which is very difficult to grow, the main reason why the disease is still widespread. Pasteur produced the virus in the brains of animals. Now scientists have taken a gene from the rabies virus and inserted it into another virus, Vaccinia, which can be used as a vaccine for rabies. They have also added to Vaccinia genes that code for proteins which stimulate the immune response in vaccinees. These enhance the vaccine's protection.

◀ A Canadian farmer administers monoclonal antibodies to a calf to prevent "scours". Prevention, rather than cure, is the basis of veterinary healthcare since animal lives are valued usually in financial terms. The cheapest and simplest solution would be a vaccine that can be added to animal water supplies.

Improving livestock

Agricultural animals are artificial beasts. Bred from elite stock selected on the basis of qualities required by humans, they are fed and cosseted and would be unlikely to survive in the wild. Modern livestock farming is already intensive. Biotechnology will intensify it even more by improving the efficiency with which animals transform their feed into useful products. As in human societies, large numbers living close together encourage the spread of infectious microorganisms. The lives of agricultural animals are short and must be productive. Infectious diseases reduce productivity by slowing growth, and treatments can be expensive relative to the value of the animal. The focus of animal healthcare, therefore, is on preventing disease by vaccination.

Animal vaccines must be cheap, easy to administer and effective. To produce a vaccine, genetic engineers move genes around in the infectious microorganism to separate the pathogenic factors (those that cause the disease) from the antigens (components that stimulate an immune response) (▶ page 91). One approach is to remove virulent genes from an infectious organism. Scours is a disease caused by *E. coli* that affects newborn calves and piglets. The resulting diarrhea and severe dehydration can kill the victims. Disease-causing strains of the bacteria produce one protein which allows the bacteria to adhere to the gut of the young animal and another which governs water loss. If the gene for the water-loss protein is removed, scours can be prevented. The organism sticks to the gut without causing diarrhea and acts as a vaccine, stimulating an immune response against the adhering protein. The vaccine is often given to pregnant animals, which pass immunity on to their offspring. A similar method has been used to produce a vaccine for leukemia in cats.

Some vaccines are made not by disarming the pathogen, but by transferring the genes coding for the antigens of pathogens into harmless organisms. This method has been used to produce a rabies vaccine. The gene for a protein on the surface of the rabies virus is inserted into the DNA of another virus, *Vaccinia*. *Vaccinia* causes cowpox but is relatively harmless. It acts as a vector, transporting the piece of rabies virus DNA to a vaccinated individual and subsequently producing an antigenic rabies protein. This stimulates a protective immune response against rabies. The vaccine is being given both to domestic dogs and wild foxes in France in an extensive program to eradicate rabies.

Subunit vaccines are produced by genetic engineering. They are purified single proteins from the surface of a pathogen which can be produced cheaply in fermentations. The great advantage of subunit vaccines is that they contain no live, potentially infectious organisms. The conventional vaccine for foot-and-mouth disease (a preparation of "killed" viruses) was responsible for nearly half of the outbreaks of the disease in Europe between 1968 and 1981 because it contained a low percentage of live organisms. It may now be replaced by a subunit vaccine. The gene for the surface protein of the foot-and-mouth disease virus has been cloned, and the protein produced in bulk in fermentations of *E. coli.* for use as a vaccine in pigs and cattle. Scientists have even identified fragments of the protein which can act alone as vaccines to prevent foot-and-mouth disease. Numerous other proteins that have potential as vaccines have also been produced in *E. coli* or in other host organisms such as yeast or mammalian cells. Virus diseases of pigs, sheep, cattle, horses, cats, dogs, mice, rats, chickens, ducks and varieties of fish may all be treated with subunit vaccines.

By keeping animals healthy, vaccines increase animal growth. Healthy animals can be made still more productive if they are given hormones to stimulate growth and the repair of tissues. Growth hormones are small proteins which can be produced by genetic engineering in microorganisms and cultured animal cells. When injected into animals, the hormones produce some startling results. Bovine growth hormone (also called bovine somatotropin or BST) increases milk yields of lactating cows by more than a quarter. Young pigs on porcine growth hormone grow faster, produce much leaner meat and use feed compounds much more efficiently. The wool of sheep treated with epidermal growth factor grows faster, and very finely, and can be removed by brushing rather than shearing. Researchers have also obtained protein hormones from chickens, salmon and goats. They are now looking for new ways of administering the hormones that avoid the need for repeated injections. The drugs cannot be given orally because proteins are digested rapidly in the stomach. Some form of slow-release device implanted in the animal is the most likely solution.

The best slow-releasing devices are the hormone-producing glands themselves. Consequently, genetic engineers are examining ways of making animals produce more growth hormones. Laboratory mice grow to nearly twice their normal size when genes for rat growth hormone are inserted into their chromosomes. Researchers have not yet been able to produce the same dramatic increase for livestock.

Genetic engineering of animals is performed on embryo cells. Only embryo cells will give rise to whole animals. Single-celled embryos, the immediate result of fusion between an egg and a sperm, are obtained either by artificial insemination or by in vitro fertilization (combining sperm and egg in the laboratory). Using microinjection, up to 30,000 identical copies of a DNA sequence are inserted into each freshly-fertilized embryo. The single cell is held still while a fine glass needle penetrates the cell and nuclear membranes. DNA is injected into the nucleus and some of it becomes inserted randomly into the chromosomes of the embryo. Subsequently it behaves as if it were the embryo's own DNA. Only a few percent of cells survive this treatment. The "transgenic" embryos are then reimplanted in the animal's ovaries to follow the normal course of development towards birth. Every cell of the new animal will contain the inserted DNA.

Genetically engineered animals offer a vast range of opportunities in agriculture. For instance, the composition of cow's milk could be altered to be more like human milk, making it suitable for feeding

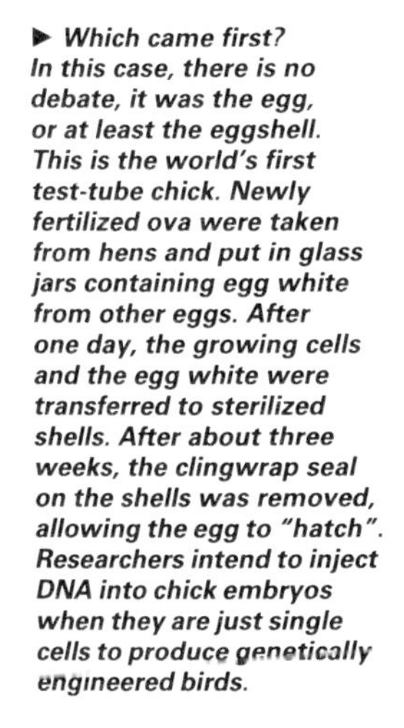

▶ Which came first? In this case, there is no debate, it was the egg, or at least the eggshell. This is the world's first test-tube chick. Newly fertilized ova were taken from hens and put in glass jars containing egg white from other eggs. After one day, the growing cells and the egg white were transferred to sterilized shells. After about three weeks, the clingwrap seal on the shells was removed, allowing the egg to "hatch". Researchers intend to inject DNA into chick embryos when they are just single cells to produce genetically engineered birds.

▼ Twin bull calves produced by splitting an embryo containing just a few cells, using a microscopic knife. Each of the halves can be subdivided. The split embryos are implanted in different surrogate mothers. Embryo splitting is a way of rapidly reproducing high-pedigree livestock. Genetically identical animals are also of interest to scientists. On the calves' backs are small pumps which inject measured doses of either a hormone or a placebo fluid into the animals.

◀ *Bovine growth hormone (BGH) increases milk yields from cows, by an average 25–30%, without the animals needing any more food. BGH is produced by genetically engineered microorganisms. In the first tests, the hormone was injected into the cows, but this is not practical for farmers. Therefore, researchers are trying to alter the gene coding for BGH by genetically engineering single-cell embryos in the hope that the resulting animals will overproduce the hormone.*

both infants and the large numbers of people who lack the full complement of enzymes needed to digest cow's milk. The high saturated fat content of whole milk might also be reduced by adding genes coding for enzymes which unsaturate the fats (though this would make butter manufacture difficult). Furthermore, fecundity, meat production and milk yields could all be increased by adding genes coding for fertility or growth hormones. The quality of other animal products such as hide, fur and wool could also be improved.

Genetic engineering is only one of the ways in which embryo manipulation can be used in agriculture. Embryo splitting, for instance, is important for producing animal clones. Embryo fusions are also possible, yielding animals with four or more parents. This technique produced the unattractive sheep-goat chimera, or "geep". The cells of chimeric embryos do not fuse and therefore the geep developed as a patch work of goat and sheep. Its coat contained both wool and mohair.

Interconvertible resources in agriculture

Biological materials are malleable. Increasingly they can be converted into one another, or at least they can be made functionally interchangeable. For instance, without manipulating genes or using any biotechnology, milk can be obtained from soybeans. The beans are harvested, crushed to release the soya oil, and purified to give a white powder. This can be reconstituted with water to give the protein equivalent of milk. Soya milk is used for cooking, for feeding infants and extensively in the food processing industry.

The British company Imperial Chemical Industries (ICI) developed the largest fermentor in the world (60 meters high and 1,500 cubic meters in capacity) to cultivate a bacterium called Methylophilus methylotrophus. Each year 150,000 tonnes of the microorganism is sold as a powdery protein-rich supplement called Pruteen for calf and chicken feed.

Soybeans are replacing cows as a source of milk. Fermentation is a substitute for soya in animal feed. Cows could be an alternative to fermentation for providing drugs for the medical profession. Milk contains a high concentration (over 10 gallons per liter) of proteins called caseins. Genetic engineers can replace the genes coding for casein with genes coding for protein drugs. Transgenic embryos are then implanted into surrogate or natural mothers, where they grow, and the animals mature as normal. Mice treated in this way produce foreign proteins in their milk. Scientists have also inserted genes for Factor IX, for the treatment of hemophilia, and alpha-antitrypsin, for lung disease, into sheep embryos. If the new ewes produce these proteins in their milk, a small herd of sheep could satisfy the world's demand for them. This production method could be 1,000 times cheaper than fermentation.

In the future, exotic spices will be produced by microbial fermentations or enzymic transformations. Natural fibers such as silk, cotton or flax may be produced by microorganisms growing in a fermentor. The fats that give the distinctive flavor and texture to chocolate may be produced in genetically engineered varieties of oilseed rape. "Mycoprotein", a mold grown in ICI's fermentor, is already replacing some of the meat we eat. It can be processed into fibrous lumps so that its texture resembles beef or lamb or chicken. It has no taste, no smell and no color. These can be provided to suit individual preferences, from other cultures of bacteria, yeasts and fungi.

For almost every biological problem, biotechnology will provide at least one solution, and usually more. The growth of wool on sheep, for instance, is limited by the amount of sulfur-rich amino acids the sheep consume. One solution being tried in Australia is to transfer a gene from peas coding for a sulfur-rich storage protein into alfalfa, a crop on which the sheep graze. Once in the stomach, however, the extra amino acids might be broken down by bacteria to produce hydrogen sulphide gas (which smells of rotten eggs) and wasteful ovine flatulence. This might be cured by genetically engineered bacteria that either would not metabolize the sulfur amino acids or would reconstitute them from hydrogen sulphide. More amino acids would thus be made available, but the sheep might not be able to absorb them quickly enough to maximize wool production. The final part of the solution, therefore, might be to genetically engineer the sheep to increase the concentration of the proteins that absorb the sulfur-rich amino acids from the stomach. All these approaches are currently being examined by research teams.

It might be most efficient to dispense with sheep altogether. Like other natural fibers, wool could conceivably be produced by genetically engineered microorganisms. Already, the genes for spider silk have been transferred to bacteria. The silk is exceptionally strong and very light. Scientists have proposed that this new material could be used for making bulletproof vests.

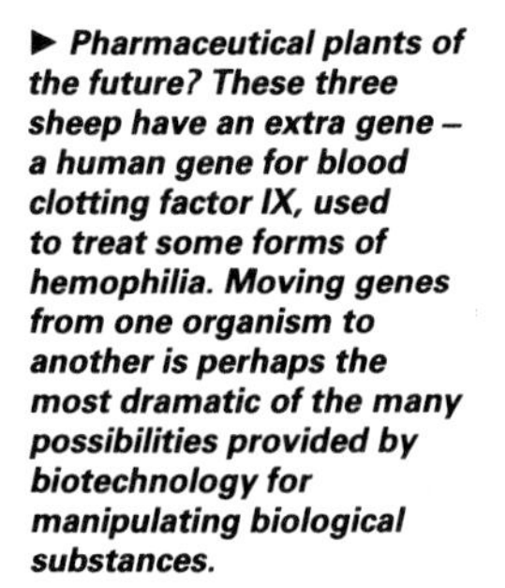

► Pharmaceutical plants of the future? These three sheep have an extra gene – a human gene for blood clotting factor IX, used to treat some forms of hemophilia. Moving genes from one organism to another is perhaps the most dramatic of the many possibilities provided by biotechnology for manipulating biological substances.

Biotechnology and Healthcare

5

How vaccines work...Safer vaccines through genetic engineering...Antio-idiotype vaccines... Contraceptive vaccines...Diet and disease... Diagnostic assays...Antibodies...DNA probe diagnosis...Genetic diseases: prenatal diagnosis... Human proteins as drugs...The artificial pancreas... Bone marrow transplants...Interleukin-2 and side effects...Somatic and germline cell gene therapies

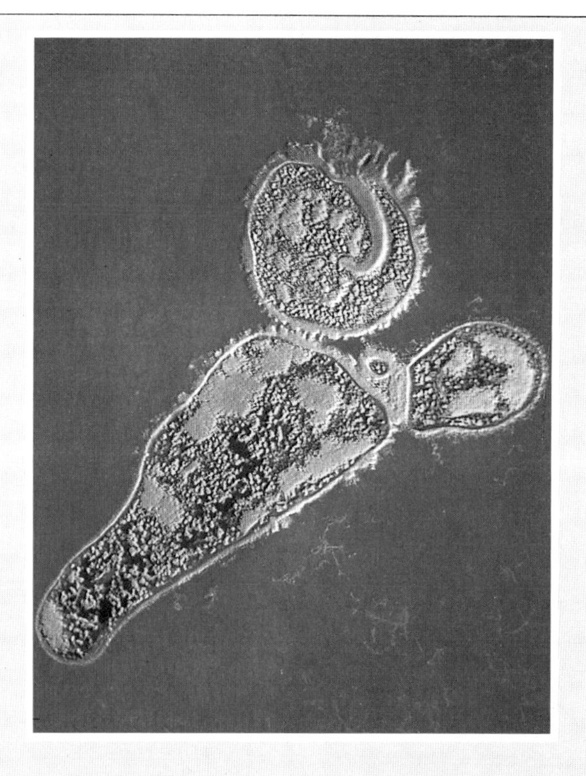

▲ The bacterium that causes whooping cough, Bordetella pertussis. Vaccinating children against the disease is a sensible precaution, though whooping cough vaccines have been associated with rare cases of childhood brain damage. High-purity vaccines from biotechnology may further reduce the very low risks.

▼ A virus related to the smallpox virus. The virus particle (yellow) is surrounded by membranes (red). In the Vaccinia virus, proteins in the membranes prime the body's immune defenses against smallpox following vaccination. With genetic engineering, new proteins can be inserted to produce new vaccines.

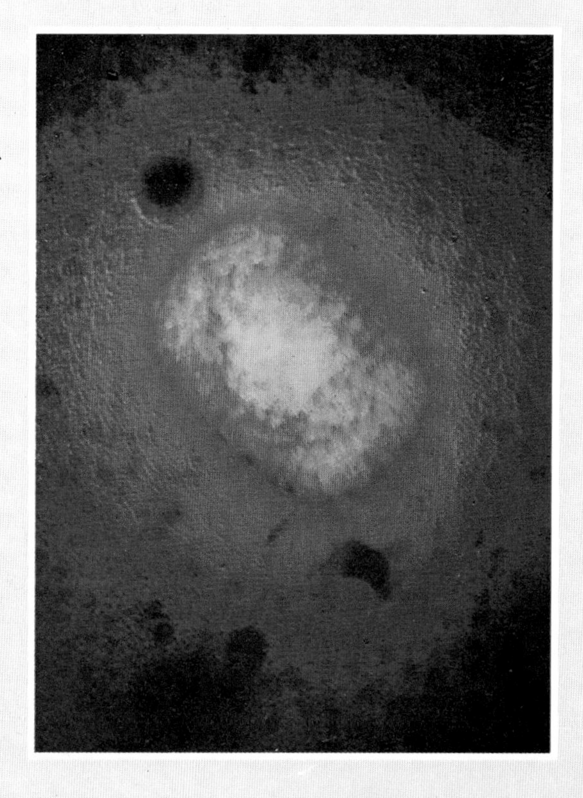

Biotechnology and healthcare

Ill health can be uncomfortable, debilitating and even degrading. If the delicate metabolic balance of the human body is upset by internal or external agents, disease may result. The three most important elements of healthcare are diagnosis, cure and prevention.

In the 18th century, the Comte de la Condamine, a French mathematician and scientist, wrote "Every tenth person and one tenth of all mankind was killed, crippled or disfigured by smallpox." As recently as 1945, most of the world's population lived in areas where the disease was endemic. In 1980 the World Health Organization (WHO) finally announced the global eradication of smallpox, the last confirmed case having been reported in 1977. The *Variola* virus, which causes smallpox now exists only in a few research laboratories.

The eradication of smallpox was made possible by a safe, simple vaccine. A vaccine is a sheep in wolf's clothing – a fairly harmless invader which resembles a pathogenic (disease-causing) organism. The body will counteract any invasion, harmful or otherwise, by producing cells and antibodies against anything that it identifies as foreign material. On a first encounter, the response takes several days to become effective, but subsequently the process is more rapid and more aggressive. The purpose of vaccination, therefore, is to make the immune system ready to counter a pathogen.

A vaccine separates the harmful effects of an infectious organism – its pathogenicity – from its antigens, the features by which the immune system recognizes the organism. Antigens are proteins, carbohydrates or other complex molecules on the organism's surface. The surface of the smallpox virus displays threads, or "tubules", of protein. Very similar proteins are also present on a closely related virus, *Vaccinia*, from which the word vaccination is derived. *Vaccinia* is a feeble form of the pathogen which to the immune system looks like the smallpox virus, but which is not virulent. When it is injected into the skin of humans, the antibodies and active cells produced give protection against *Variola* to prevent smallpox. Vaccinia is an example of a live vaccine and, as such, simulates the disease particularly well. Only small amounts of the virus are needed because it multiplies for a few days in the skin until the immune response is roused.

Another way of separating pathogenicity and antigenicity is to kill the organisms without altering their antigens. For instance, the basis of the whooping-cough vaccine is a preparation of killed *Bordetella pertussis*, the bacterium that causes the disease. It is usually administered as a triple vaccine, a preparation also containing diphtheria and tetanus toxoids. Toxoids are altered forms of the toxic proteins which help the microorganisms responsible for diphtheria and tetanus to cause the diseases. Several booster doses of "killed" vaccines may be needed to build up full immunity.

New vaccines are needed. Genetic engineering may help reduce the risk (which is extremely low already) of virulent organisms being present in vaccines. It may even provide vaccines against diseases such as AIDS and malaria, where none exist at present.

Producing vaccines requires stringent safety measures because it involves growing large amounts of potentially harmful organisms to obtain the antigens. Luckily, many important antigens are proteins and can be made in harmless organisms through genetic engineering. Genetic engineering also provides a way of producing vaccines even if the infectious organism cannot be grown. For instance, the virus that causes hepatitis B, a liver disease, cannot be grown in animals or in artificial culture. So genetic engineers have taken the gene that codes for a surface protein (hepatitis B surface antigen – HBSA) and inserted it into *E. coli*, or into yeast. When the proteins are produced in the new hosts, they are easy to purify for use as vaccines. It is not even

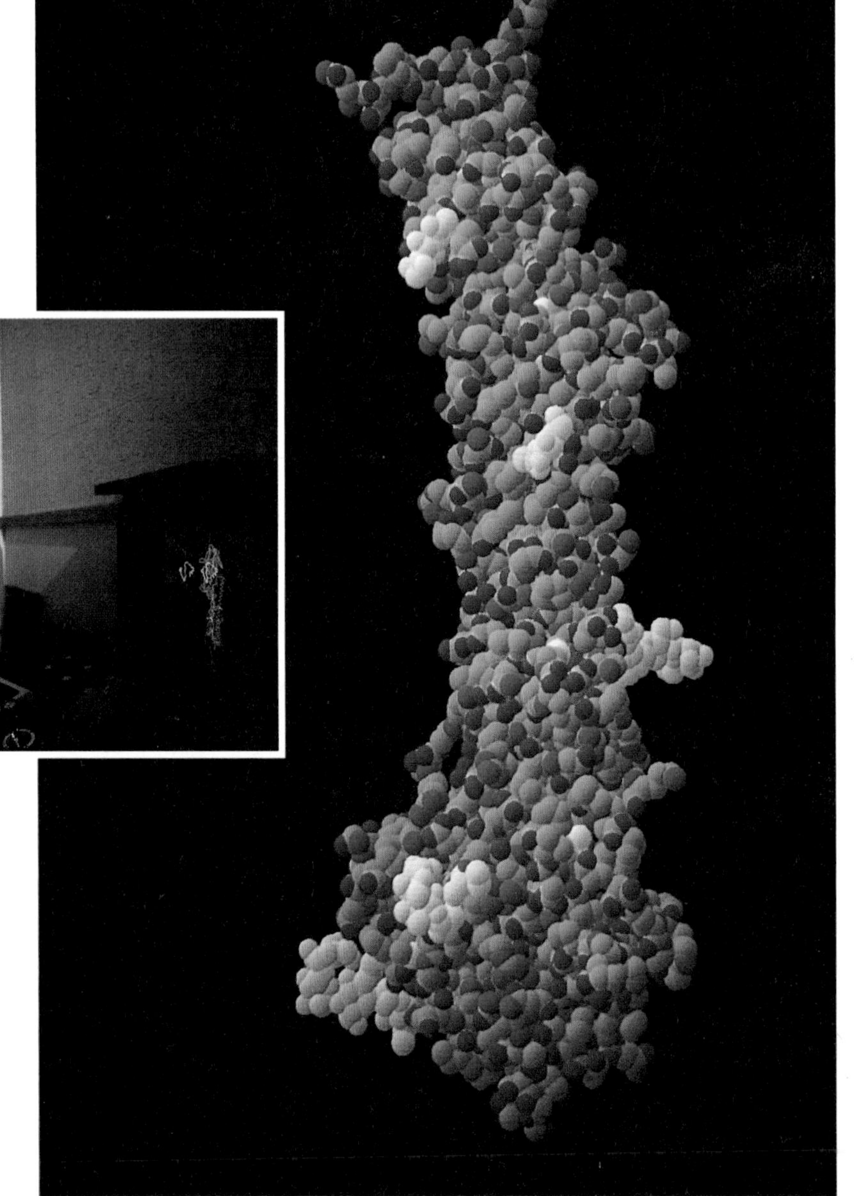

▼► Scientists know the structure of the hemagglutinin protein from the 'flu virus. They have found that the body's immune system acts against only certain parts of the protein, usually the most protuberant parts of the molecule (white). The different colored parts of the hemagglutinin molecule being studied on the computer screen have been tested individually as potential vaccines.

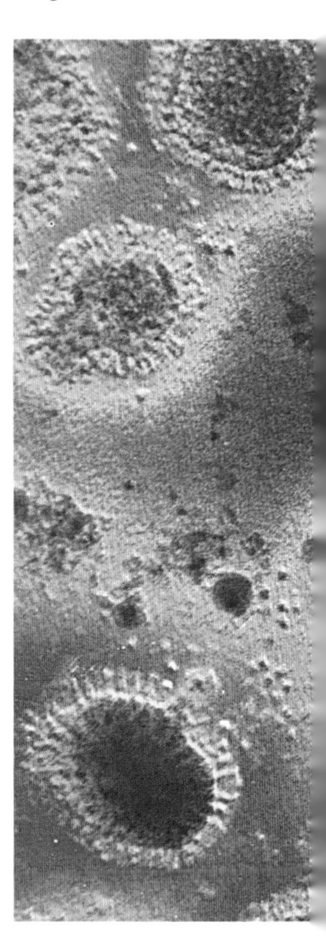

► Hepatitis B virus, which causes acute inflammation of the liver. Its particles are made up of a few different types of protein and DNA. The gene coding for the protein at the virus's surface has been transferred to yeast cells. When the protein is produced there, it forms particles which are very like the virus but contain no DNA. They are, therefore, noninfectious, and are being used as vaccines.

necessary to produce the whole protein to obtain a genetically engineered vaccine. Only small regions of the protein, called epitopes, are bound by antibodies (proteins produced as part of the body's immune system). Epitopes may contain only a few amino acids and, because the protein turns and twists back on itself, these are not necessarily linked together directly. Scientists have reasoned that short amino-acid chains could mimic the epitopes of antigens and be used as vaccines. For instance, on the surface of the human immunodeficiency virus (HIV) – the causative agent of AIDS – there are several proteins which can elicit an immune response. Researchers have found that short regions of the proteins produced in *E. coli* and then joined to carrier proteins elicit the same sort of immune response in laboratory animals as does the AIDS virus itself. The work to produce a vaccine against AIDS still has a long way to go. Indeed, it is not clear whether any vaccine would be effective in limiting the spread of disease.

Jabs against influenza, on the other hand, have already been used widely. Immunity to 'flu only lasts a year or so, not because the vaccines are no good, but because the virus keeps changing its protein coat. Scientists in Israel have investigated several strains of the 'flu virus, looking for regions that did not vary (perhaps because they were essential to the virus) but were exposed sufficiently to stimulate an immune reaction. They found a region of 18 amino acids, which they knew, from protein structure studies, was located in an exposed part of a viral protein called hemagglutinin. This region was short enough to be produced chemically in a protein synthesizer. Joining this synthetic peptide to a carrier protein produced a protein that protected mice from several different strains of the influenza virus.

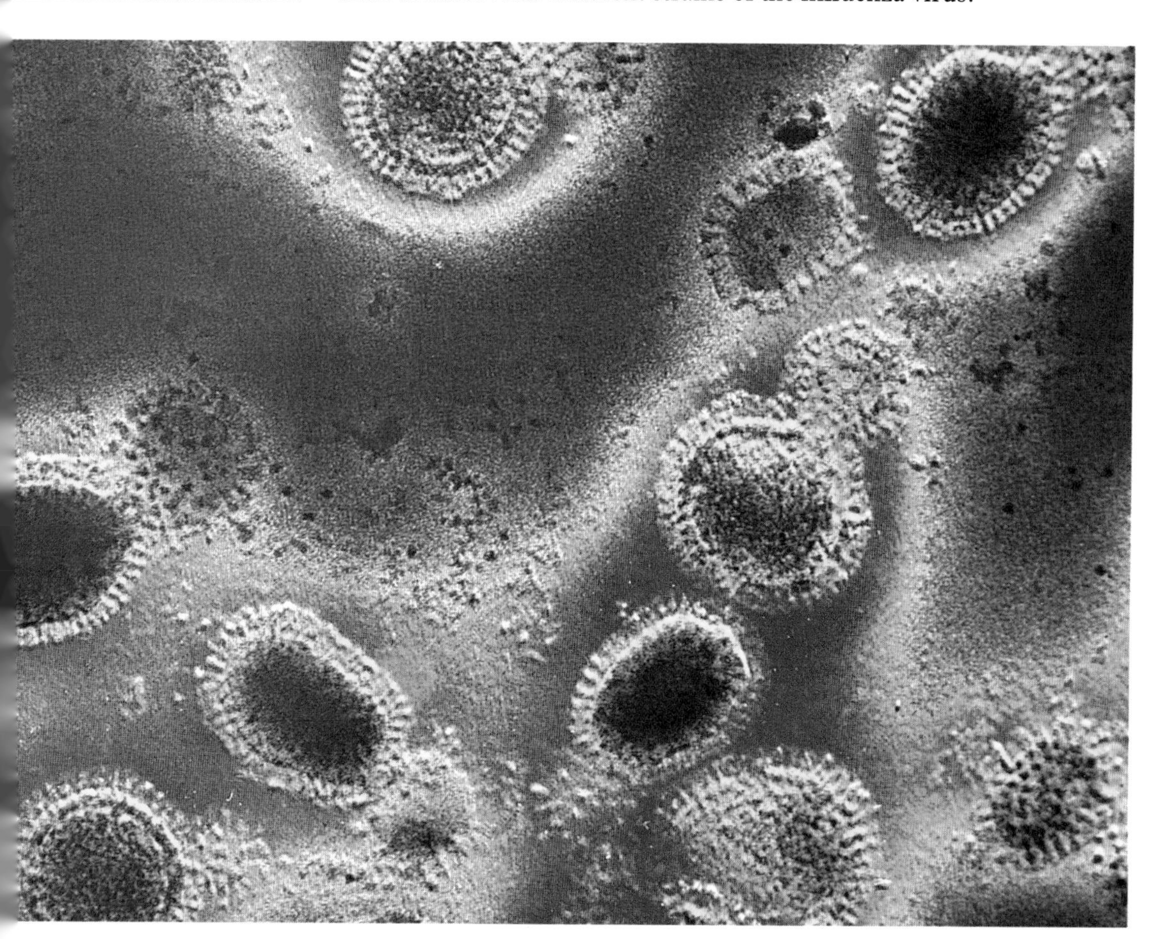

◄ The influenza virus consists of a core of protein and RNA surrounded by a membrane with spikes embedded in it. These are the hemagglutinin protein, seen in detail here. 'Flu comes in epidemics because the virus alters its hemagglutinin. This nullifies the protection against infection developed by the body's immune system. Scientists are trying to identify those parts of the hemagglutinin molecule that do not change from one wave of 'flu infections to the next.

◀ Homo sapiens – polluting the planet with people. The 19th century British economist Thomas Malthus believed that famine would inevitably limit the expansion of the human population. Malthus was wrong. Although there is still famine, improvements in agriculture have allowed the world's population to expand dramatically.

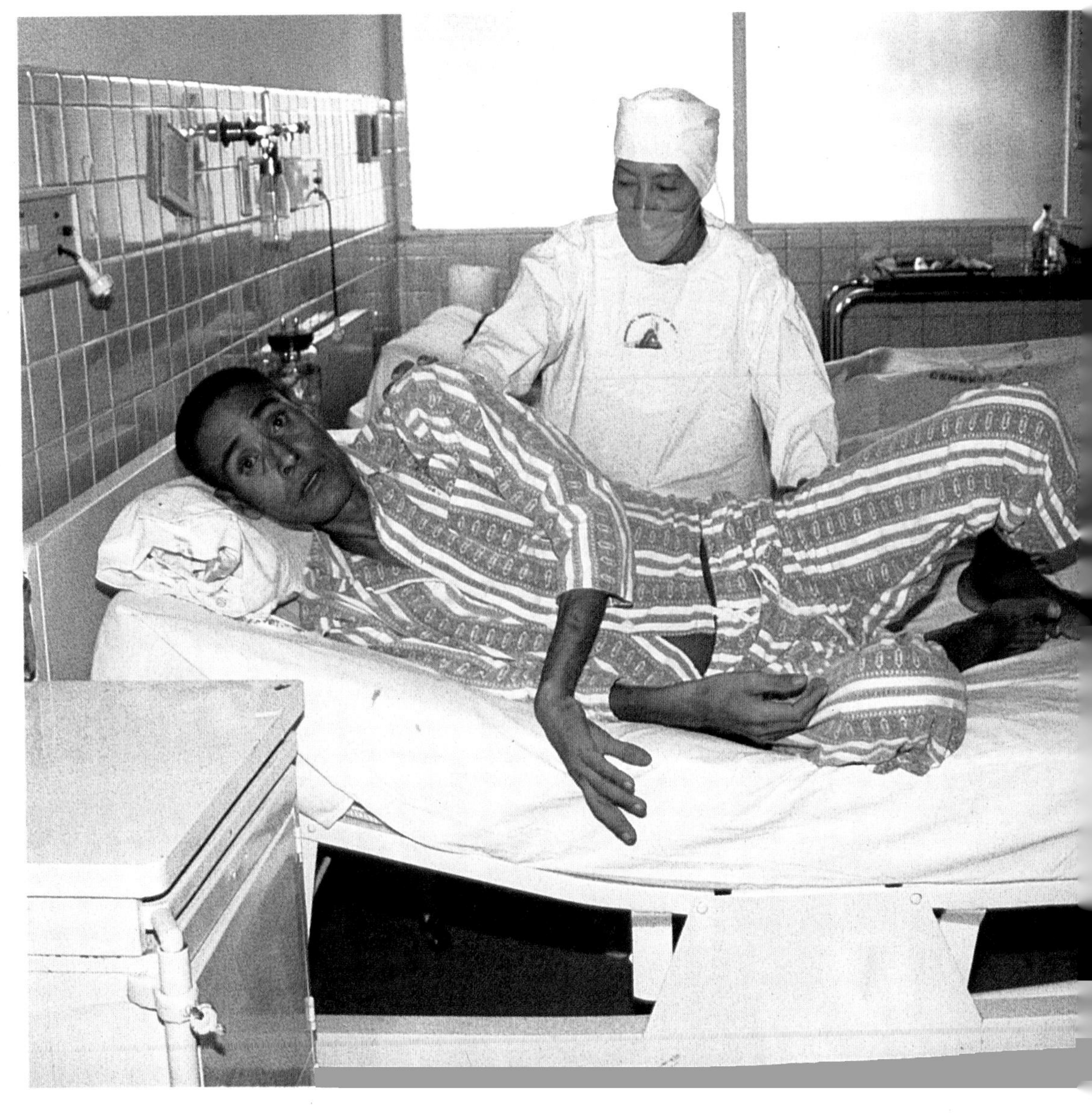

▶ An AIDS patient in a Mexican hospital. AIDS has been nurtured by the modern world. The ease of air travel and a more relaxed attitude to social relations have accelerated the global spread of AIDS. So too has international trade in transfused blood. Because AIDS affects the developed countries, they have invested lots of money in researching it, unlike parasite diseases such as malaria or sleeping sickness, which kill many more people each year.

Protein and peptide preparations are "dead" vaccines. "Live" vaccines can also be produced by genetic engineering. If genes from other organisms are inserted into the DNA of *Vaccinia*, the virus used as a vaccine will produce the corresponding proteins as it grows inside human cells. Experimental vaccines to protect animals against infection from rabies, herpes, hepatitis B, malaria and influenza have been produced in this way. A similar vaccine is being developed for AIDS. Up to 25 extra genes can be inserted in *Vaccinia* DNA and there are plans to produce multiple vaccines to confer immunity to several diseases simultaneously. Scientists do not know how the immune system will react. Biotechnologists are also developing vaccines based on antibodies. Special antibodies (anti-idiotype antibodies) can mimic the three-dimensional shape and distribution of surface properties of antigens. How can antibodies, which are usually produced in response to vaccines, themselves be vaccines?

A hand thrust into a bucket of modeling clay will leave a distinct impression. Wet plaster poured into this "mold" and left to dry will then create a plaster copy of the hand. Although the cast is not the real hand, it would leave an identical impression to the hand in a second bucket of clay. Antibody-based vaccines work on the same principle. An antigen will engender the production of antibodies whose binding sites complement its shape and other properties. Like any proteins, these antibodies can, in turn, act as antigens, to induce the formation of a second set of antibodies. Antibodies directed against the antigen-binding site of the first antibody (anti-idiotype antibodies) will have properties in common with the original antigen, including the ability to act like a vaccine. Anti-idiotype vaccine will be important either when the amino acid sequence of an antigen is unknown (or the antigen is not a protein) or if the antigen is scarce. Furthermore, since they are far removed from the original pathogen, anti-idiotype vaccines should be safe.

▼ *One way to slow down the rapid rise in the human population is to improve birth control. These devices, when implanted in women, slowly release hormones into the body. They are designed to be effective for five years and are most likely to be used in countries where regular access to other methods may be difficult.*

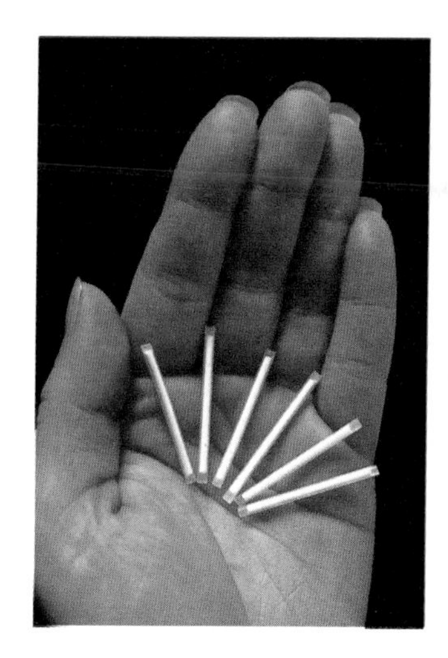

Population changes and healthcare

One of the most serious dangers to the world's health is the spread of the organism *Homo sapiens*. Sixty years ago there were two billion human beings on Earth. In 1987, the figure passed five billion and it is expected to reach six billion before the end of the century. Every minute, over 150 people are added to the tally. Populations grow when births exceed deaths. On a global scale, the birth rate has been steady for several hundred years, at around 27 births per year per 1,000 people. At the same time, the average death rate has fallen (partly because of improved vaccination) and now only 11 people per year die out of every 1,000. In advanced countries, a wider use of contraception has largely restored the balance between birth and death rates.

In developing countries, however, populations are doubling, on average, every 34 years. Where standards of health education are low and medical services poor, many contraceptive methods used in the West are inappropriate. Thus WHO has concentrated its research on the development of a contraceptive vaccine. Vaccination is familiar to people in developing countries, its effects would be long-lasting but reversible, and it would be cheap and easy to administer. The vaccine, which is still in the experimental stage, is directed against the hormone human chorionic gonadotrophin (hCG), which appears in the blood only during pregnancy. It is produced by the embryo to help sustain early pregnancy. The vaccine is being tried in India on women volunteers who have already had children.

Diet and disease

Malnutrition and starvation in overpopulated countries might be overcome in the short term by intensive agriculture and the clearance of vast areas of land for intensive crop cultivation. Biotechnology will help to increase agricultural productivity, providing nitrogen-fixing *Rhizobium* bacteria with which farmers can inoculate their crops, improving livestock and producing drought-resistant plants. Farmers in developing countries could also grow more nutritious crops with the help of genetic engineering. For instance, researchers at the International Potato Center in Peru have inserted genes coding for proteins high in essential amino acids (those that the human body cannot make itself) into potatoes, a staple crop in parts of South America.

▼ *Many children still enjoy sugary drinks. In countries where there is an excess of food, people easily develop bad eating habits early in life. However, some firms are now producing foods with lower fat and sugar contents. These may improve dietary awareness and help reduce the incidence of obesity, heart disease and strokes in later life.*

People in developed nations are now beginning to alter their diet for health reasons. However, eating habits die hard, so there is a growing demand for "diet products" which look, smell and taste just like traditional foods but have fewer calories. Sugar in soft drinks, in puddings and in processed food is being replaced with biotechnological sweeteners like aspartame and thaumatin. Likewise, the fatty ingredients of food are being substituted with "Cal-0 fats" – fats that have been processed by microorganisms, but are not absorbed from the food and so have no calorific value.

Diagnosis of diseases depends on information passing from patient to doctor. Sick people may complain of feeling "under the weather", "out of sorts", "off color" and so on. To a doctor the patient's description may indicate the cause of the complaint. Some diseases, however, can only be diagnosed by chemical or biochemical tests. For instance, one person in every 10,000 is born with a genetic condition called phenylketonuria, which, if it goes unnoticed, can lead to severe mental impairment. However, the disease can be prevented by putting the patient on a special diet that is low in the amino acid phenylalanine. In many countries, newborn children are subjected to a simple biochemical test for phenylpyruvate, a toxic substance that forms in the blood of sufferers.

► *In future, researchers may be able to design new drugs. The protein thermolysin has a similar structure to enzymes that cause hypertension. The white structure, hydroxamine, stops thermolysin working by occupying the enzyme's binding site. Researchers ask the computer to alter amino acids of thermolysin so that the molecule resembles the disease-causing enzymes. From this, they can try to work out how to change hydroxamine so that it inhibits those enzymes.*

To penetrate the biochemical secrets of disease, scientists have designed three main categories of diagnostic aids. These are enzyme assays, antibody assays and DNA probe assays. In all three, biological polymers (proteins and DNA) are used to provide information on the molecular symptoms of disease.

Enzyme tests usually give information for the detection of small molecules like sugars. Diabetics, for instance, use a simple biochemical test for measuring their glucose level to help them decide whether to take more insulin. The test enzyme, glucose oxidase, produces peroxide (the chemical used to bleach hair) in response to any glucose in a sample drop of blood. A second enzyme converts the peroxide produced by glucose oxidase to water and simultaneously changes TMB, another chemical present, into a deeply colored dye. The amount of dye formed corresponds to the amount of peroxide, and hence of glucose, in the sample.

The enzyme reactions can be performed in liquid solutions in test-tubes, but for convenience the enzymes are often attached to a solid surface. Solutions of enzymes and other reactants might, for instance, be dried onto strips of cellulose fiber paper or a swab. When needed, a strip or swab can be dipped into the sample of blood or urine (or a drop of fluid can be placed on it). In this form, the tests last a long time without "going off" and can be used by nonmedical personnel including patients.

► *Technicians at the International Potato Center in Peru prepare new strains of potato for storage. Gathering a wide range of different types of plants is the first step in breeding new, improved staple crops in developing countries. At the same time, genetic engineering work is helping to improve the quality of crops more rapidly.*

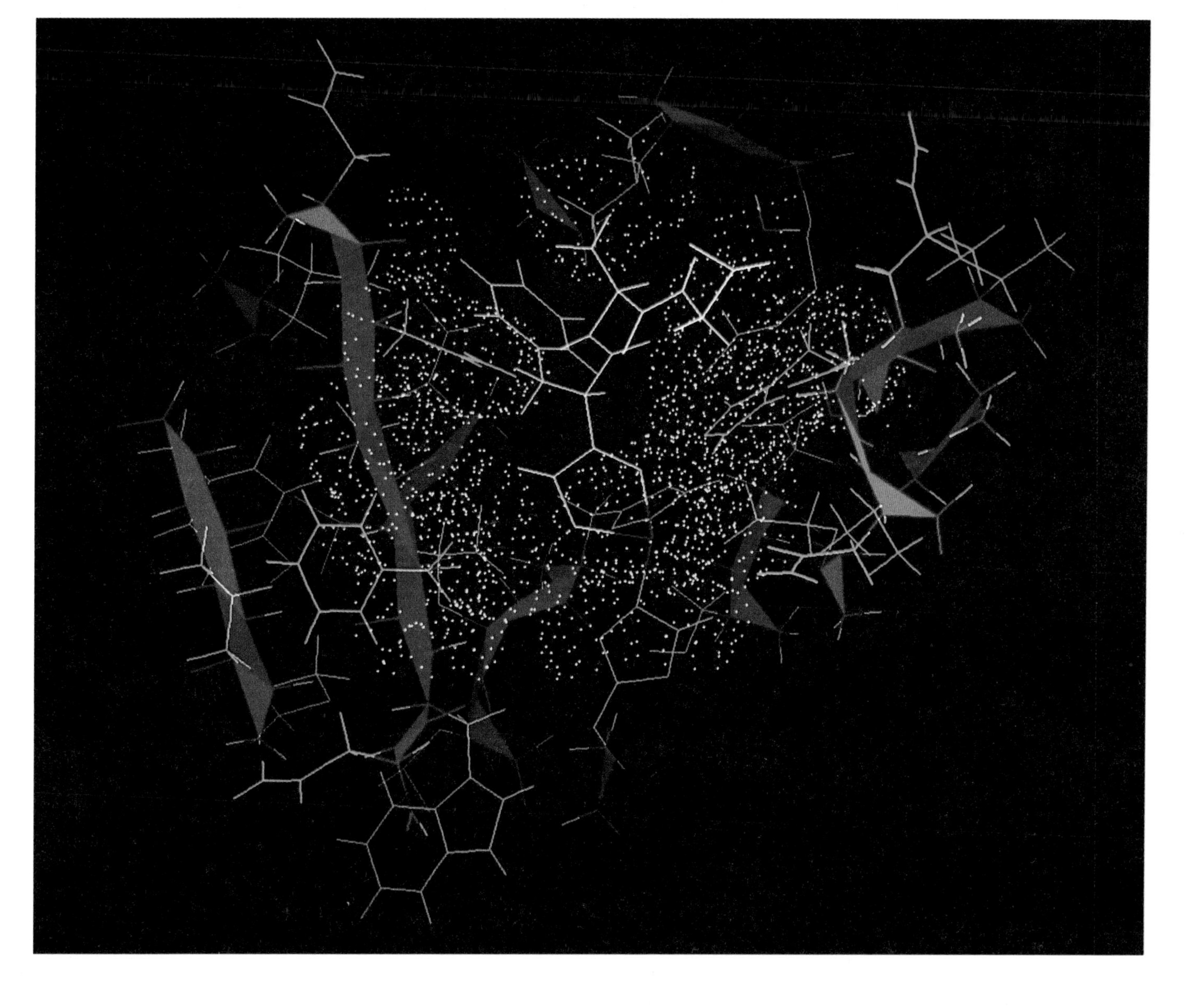

Antibody assay

There are thousands of enzymes but few are used in diagnostic tests. Antibodies are much more widely employed. At one time, sheep, rabbits and horses were the main sources of antibodies. However, the quality of such preparations is difficult to maintain, especially since no two animals will ever produce the same set of antibodies. The discovery of hybridoma cells in 1975, offered a way of making the production of antibodies completely consistent. Each cell can produce unlimited amounts of monoclonal antibodies with a single specificity.

Their exceptional capacity for specific binding of biological molecules makes antibodies (and particularly monoclonal antibodies) powerful "homing" devices. However, when antibodies bind antigens, neither component changes in any detectable way. In diagnostic assays, signals such as radioactive or fluorescent compounds need to be attached to antibodies. When radioactive antibodies bind an antigen, the antigen effectively becomes radioactive and easily detectable. Radioactive labeling, however, requires special safety measures and is not always convenient. Other labels are now being used instead, notably enzymes. For instance, the peroxidase enzyme can be chemically linked to antibodies to combine biological specificity with biological signal generation. Light-producing enzymes are also used. Enzymes and other labels are often attached indirectly to the antibodies by linking molecules. Biotin, a vitamin, and avidin, a protein that binds biotin, provide a kind of universal link. With biotin attached to antibodies, any one of a number of labels (enzymes, fluorescent compounds, etc.) bonded to avidin can be used in the assay.

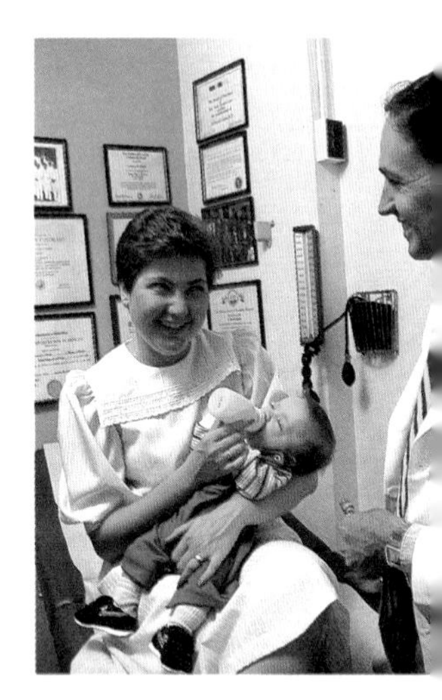

▲ ***Monoclonal antibodies can be used to treat disease, and monitor the progress of recovery. Soon after this young mother had a kidney transplant operation, monoclonal antibodies were applied with other agents to prevent organ rejection. Subsequently, her blood was sampled regularly and tested with the same monoclonal antibodies in order to detect signs of transplant rejection.***

Pathogenic microorganisms can be detected in an impure clinical sample such as a skin swab, pus or feces. If a microscope slide on which the specimen has been smeared is flooded with a solution containing fluorescence-labeled antibodies against the pathogen, the organisms will show up as bright spots.

Many antibody assays start by attaching the antigen to a surface. In enzyme-linked immunosorbent assay (ELISA for short), for instance, antibodies bonded to the inside of small test-tubes effectively extract the antigen from solutions placed in the tubes. In the next stage, a second antibody with an enzyme attached binds to the antigen. The third stage of ELISA is the production of color change reaction by the bound enzyme.

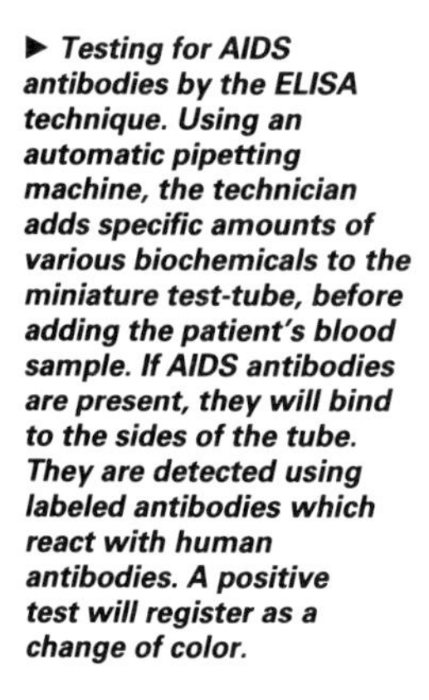

► ***Testing for AIDS antibodies by the ELISA technique. Using an automatic pipetting machine, the technician adds specific amounts of various biochemicals to the miniature test-tube, before adding the patient's blood sample. If AIDS antibodies are present, they will bind to the sides of the tube. They are detected using labeled antibodies which react with human antibodies. A positive test will register as a change of color.***

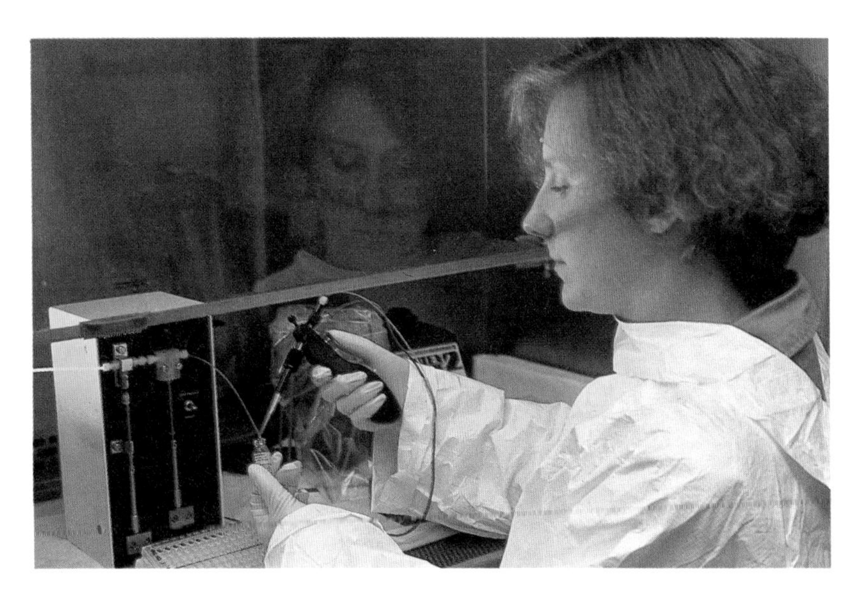

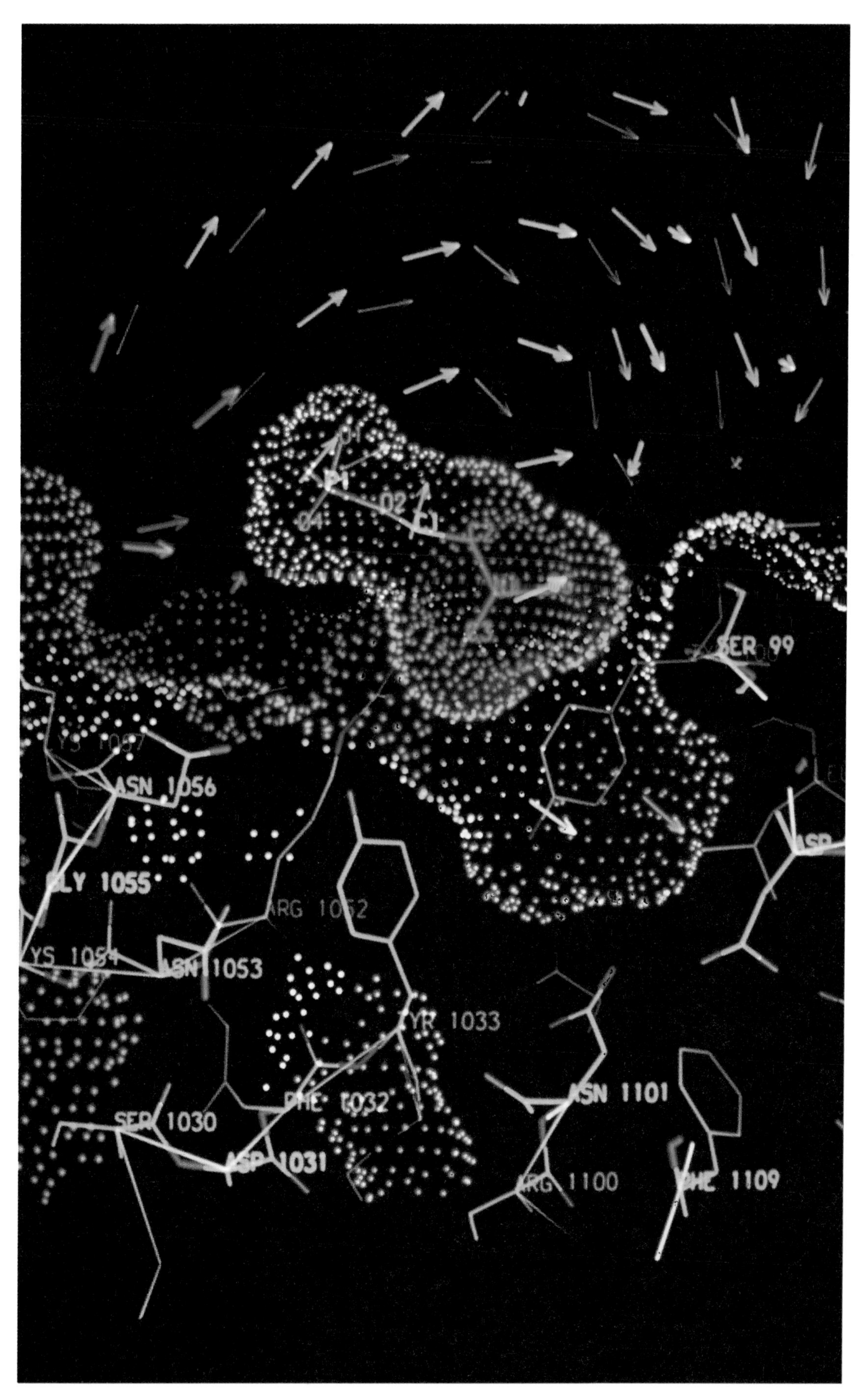

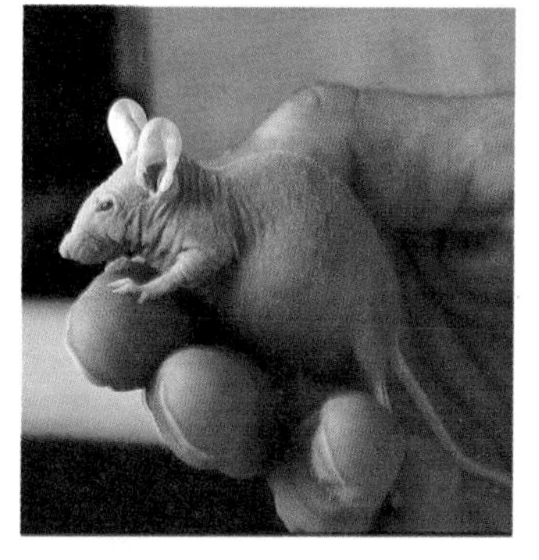

◄▲ *Hybridoma cells implanted in nude mice grow as tumors and produce monoclonal antibodies in high concentrations. Scientists harvest these and purify them to be used, for instance, in diagnostic kits for infectious diseases. Antibodies bind specifically to one particular substance (left). The antibody binding site (bottom) is complementary to the antigen (center; a substance called choline) in both shape and electric charge (red or blue dots).*

Ricin, used in cancer therapy, is so toxic that a single molecule of it is enough to kill a human cell

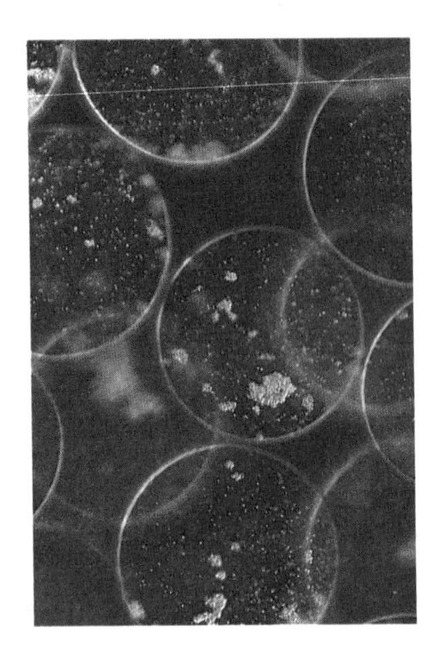

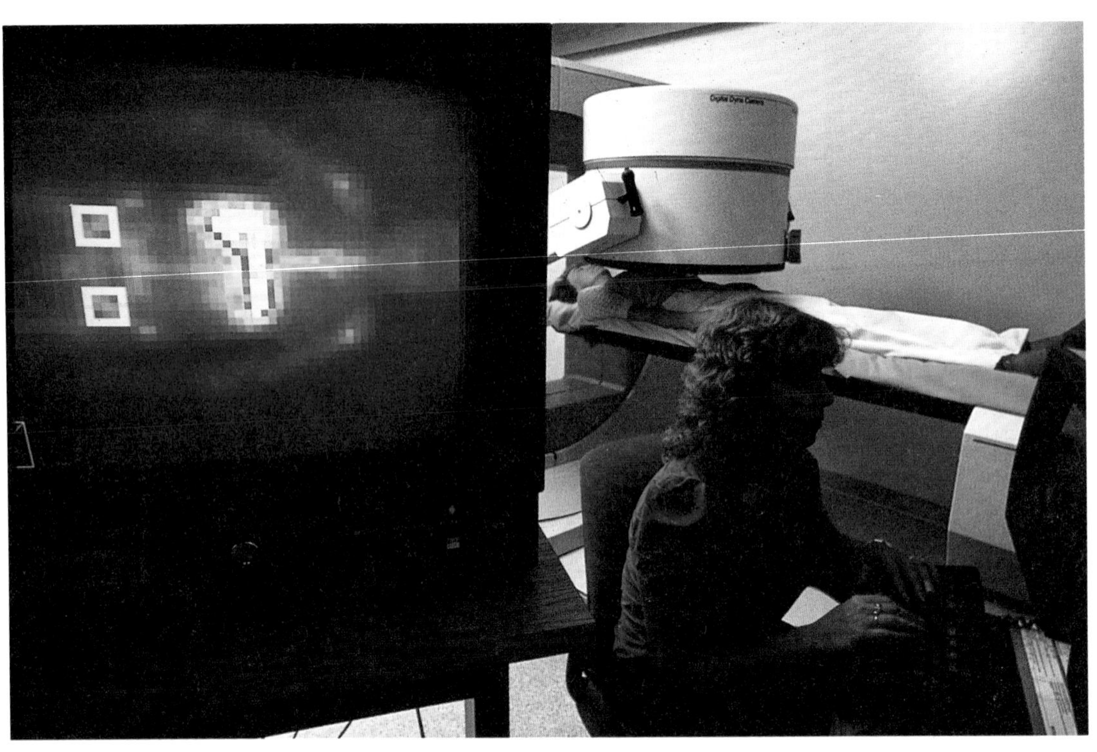

▲ *A biocompatible membrane in the capsules protects hybridoma cells from buffeting while allowing nutrients to diffuse inside. The cells grow and produce monoclonal antibodies, which are retained by the capsule and become more concentrated. This makes it easier to purify them.*

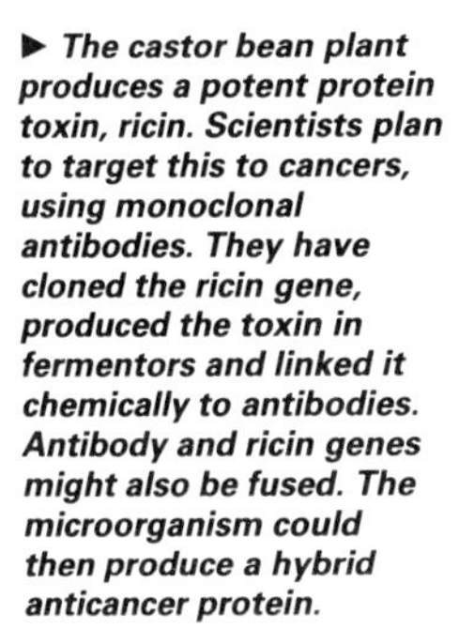

► *The castor bean plant produces a potent protein toxin, ricin. Scientists plan to target this to cancers, using monoclonal antibodies. They have cloned the ricin gene, produced the toxin in fermentors and linked it chemically to antibodies. Antibody and ricin genes might also be fused. The microorganism could then produce a hybrid anticancer protein.*

When macromolecules like proteins or carbohydrates form part of the surface of cells, they can be used to identify the cell. The *Salmonella* bacteria, for example, which sometimes cause food poisoning, are distinguished by antibody-based tests that identify their carbohydrate coat, a protein appendage called a flagellum (which is used for swimming), and other antigens responsible for virulence. These characteristics are equivalent to eyewitness descriptions of criminals. No single antigen is definitive, just as many people have buck teeth or red hair or blue eyes. A combination of all three, however, can be compelling evidence. Microorganisms also leave molecular fingerprints – single antigens which identify the organism. Infected cells of AIDS sufferers, for instance, contain proteins produced by the AIDS virus. Cancers, too, can be identified by their associated antigens. Patients with colorectal, lung, or breast cancer have raised levels of carcinoembryonic antigen (CEA). In cancer and infectious diseases, antibody tests can be used not only to diagnose the disease initially, but also to monitor its progress during treatment.

As an alternative to test-tube methods of diagnosis with antibodies, several research companies are trying to incorporate antibodies into miniature sensors on silicon chips or fiber optic devices based on laser technology. Furthermore, if antibodies that bind tumors are labeled with metallic or radioactive compounds and injected into patients, doctors can use X-rays to detect the tumors and measure their positions.

Beyond diagnosis, the same monoclonal antibodies might even be used therapeutically. When antibodies bind antigens, they provoke aggressive action against the antigen from cells of the immune system. If this fails to have any effect on the tumor, the antibodies could act as a guidance system for cell-killing drugs. Extremely poisonous substances could be acceptable as drugs if targeted specifically to tumors by antibodies. Such substances, called immunotoxins, consist of antibody molecules linked to very toxic proteins. One of these proteins, ricin, which is extracted from beans of the castor oil plant, is so poisonous that one molecule is enough to cause a cell to die. Immunotoxins are like guided missiles. They work by the antibody molecule targeting the toxin specifically to cancer cells.

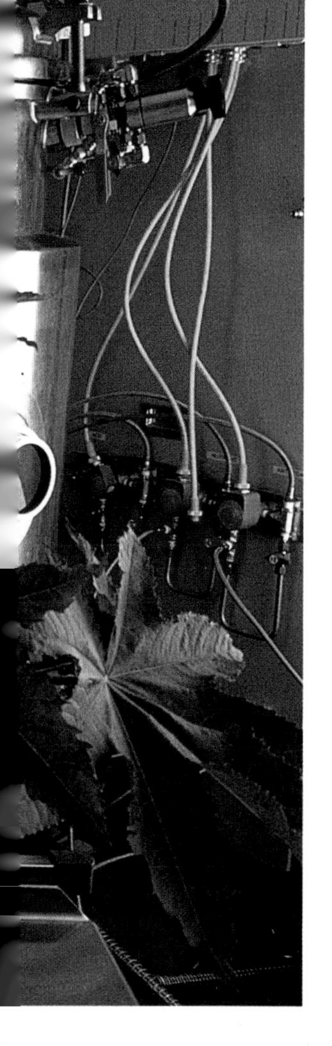

◀ To find the cancer, doctors first inject the patient with monoclonal antibodies which bind to cancer cells. The antibodies have cobalt metal attached to them. With the aid of a body scanner which detects cobalt, doctors can assess the location and size of the main tumor as well as the full extent of its spread through the body.

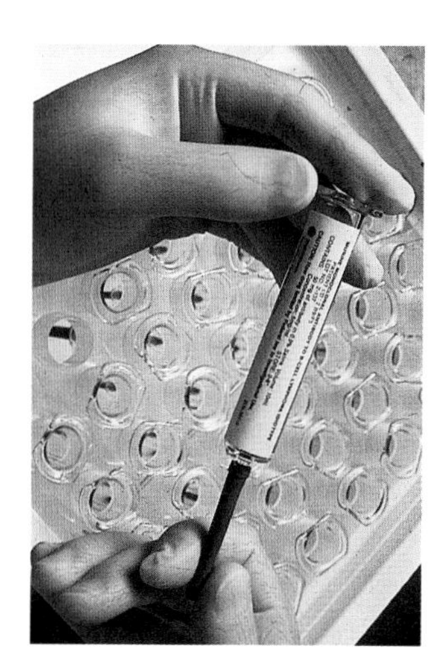

▼ Vials of monoclonal antibodies. When antibodies bind to cancer cells, they cause the body's immune defenses to attack the tumor. Monoclonal antibodies used in therapy must be extremely pure. All DNA has to be removed to ensure that cancer genes from the hybridoma cells are not injected into patients.

DNA probes

As their name suggests, DNA probes take diagnosis of disease beyond the surface of cells and their products, to the genetic material. Proteins and other metabolites are only manifestations of genetic composition and so may not always provide the best sort of diagnosis. Analysis of DNA, since it deals directly with the genes, is the most intimate and the most irrefutable form of diagnosis.

DNA probes have been used since the 1970s for diagnosing infectious diseases. Although antibody-based assays are usually quicker and easier to perform, they are often inappropriate for microorganisms that cannot be grown in the laboratory or that vary their antigens. DNA probes have successfully diagnosed glandular fever, hepatitis B, malaria, Legionnaire's disease, bubonic plague and venereal diseases such as AIDS, herpes, gonorrhea and chlamydia infections. In diagnosing infectious diseases, it is often important to be able to distinguish pathogenic strains of an organism from relatively harmless ones. Hence, DNA probes are often directed against genes that are associated with virulence – such as those coding for toxins or antibiotic resistance.

▶ *An X-linked genetic disease, hemophilia (absence of blood clotting) in the royal houses of Europe. Queen Victoria was a carrier of hemophilia and passed the defective gene to her daughter Beatrice (fourth from right). Two of Beatrice's sons, Leopold and Maurice (extreme left and right) were hemophiliacs. Their sister, Victoria Eugenie (third from right) was a carrier. Women are carriers if one of their two X-chromosomes bears the genetic defect. Men who have the genetic defect on their one X-chromosome inherit the disease. Two of Victoria's five daughters were carriers.*

Genetic disease

Most of the excitement surrounding DNA probes stems from their use in diagnosing, not infectious disease, but inherited, or genetic, disease. There are over 3,000 diseases that are due to defects in single genes. Most have no cure, and their causes are unknown. Using a DNA probe to track down a defective gene, it is possible to diagnose many of these conditions. Scientists group genetic diseases in three classes according to the location of the gene and its behavior. These are sex-linked, autosomal recessive and autosomal dominant.

Most sex-linked diseases are X-linked. In X-linked diseases, males with the defective gene will have the disease, females with it will be carriers. Women who are carriers of X-linked conditions have a 50 percent chance of passing the disease to their sons or the carrier state to their daughters. One of the most tragic X-linked conditions is Duchenne muscular dystrophy. Apparently normal babies develop with only the usual childhood problems until the age of three or so. Then, gradually, they start to become unsteady on their feet, losing their strength and control of their muscles. Confined to a wheelchair by 10, they are usually dead before their teens are out.

The autosomes are nonsex chromosomes. In autosomal recessive diseases there are many more carriers of the disease than sufferers. For instance, 1 in 20 white people is a carrier of cystic fibrosis but only 1 in 1,600 gets the disease. This is because the normal gene prevents the effects of the mutant form. It takes two defective copies of the gene,

◀ *Woody Guthrie, the dustbowl folk singer, died in a mental institution from Huntington's disease. His mother and many of his children died in the same way. In Huntington's, a defective gene on a nonsex chromosome (autosome) dominates the healthy gene. Everyone who inherits the bad gene gets the disease. The tragedy of Huntington's is that the disease's symptoms do not appear until middle age, when sufferers may already have passed the gene to their children.*

▲ *Red blood cells in sickle-cell anemia. A genetic mutation produces defective hemoglobin and distorts the blood cells. The disease can be diagnosed with the RFLP technique, which detects changes in single bases of DNA. Often, scientists trace the pattern of single base differences through a family to diagnose genetic disease.*

one from each parent, before the disease appears. With autosomal dominant diseases, however, all carriers are sufferers.

In diagnosing genetic disease, doctors attempt to trace the inheritance of a disease gene through a family tree. Finding out which gene (healthy or defective) has been inherited is not simple, because one person's chromosomes look identical to another's. However, there are differences, or polymorphisms (Greek, "many forms") in DNA sequences littered randomly throughout the chromosome. These polymorphisms arose by single base changes many thousands of years ago and are now often spread evenly in the population. There is a good chance that a patient's DNA contains a distinctive combination of these markers. Restriction fragment length polymorphisms (RFLPs) are a special group of polymorphisms which occur at the cut sites of restriction enzymes. Restriction enzymes will cut a given piece of DNA into a distinctive set of fragments which can be separated electrically to form a pattern of bands, like a bar code. Scientists compare the child's DNA with that of its normal and affected relations to work out which cutting sites are associated with normal genes and which with mutant types. DNA from parents, grandparents, aunts, uncles, brothers and sisters will all be examined to build up a pattern of the disease. If the child's DNA shows the same restriction-enzyme bar code as affected relatives, the chances are that the child has inherited the mutant gene. Cutting sites vary from family to family, so there is no universal test.

Prenatal diagnosis

Perhaps the most important feature of DNA probe diagnosis is that it can be performed before birth. This is not entirely new. Women over 35, who are at relatively high risk of giving birth to children with Down's syndrome, often undergo amniocentesis, a harmless process for obtaining cells from the fetus. If chromosome abnormalities are apparent, a sure sign of the syndrome, the woman can then decide whether she wants the pregnancy to continue. In families with a history of Duchenne muscular dystrophy, the same test has been used to assess the sex of the child. If it is a girl, it will not be born with the disease (although it may be a carrier). A boy, however, would have an even chance of being affected. Whether or not to abort would then be the choice facing the parents. At the late stage at which amniocentesis is performed, however, abortion holds its own risks, not the least of which is the psychological effect it can have on the mother.

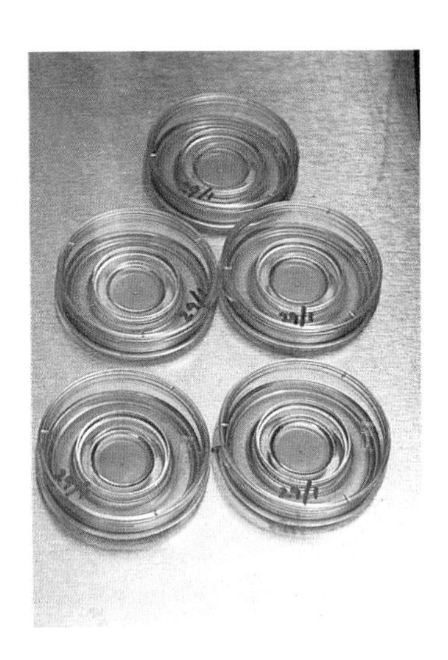

▲ *Petri dishes containing human ova await the addition of sperm to complete in vitro fertilization. Prospective mothers are given drugs to induce superovulation and then the eggs are removed surgically or flushed gently from the womb. After fertilization, eggs can be implanted in the mother or frozen for later use. Genetic tasks might be performed on fertilized embryos.*

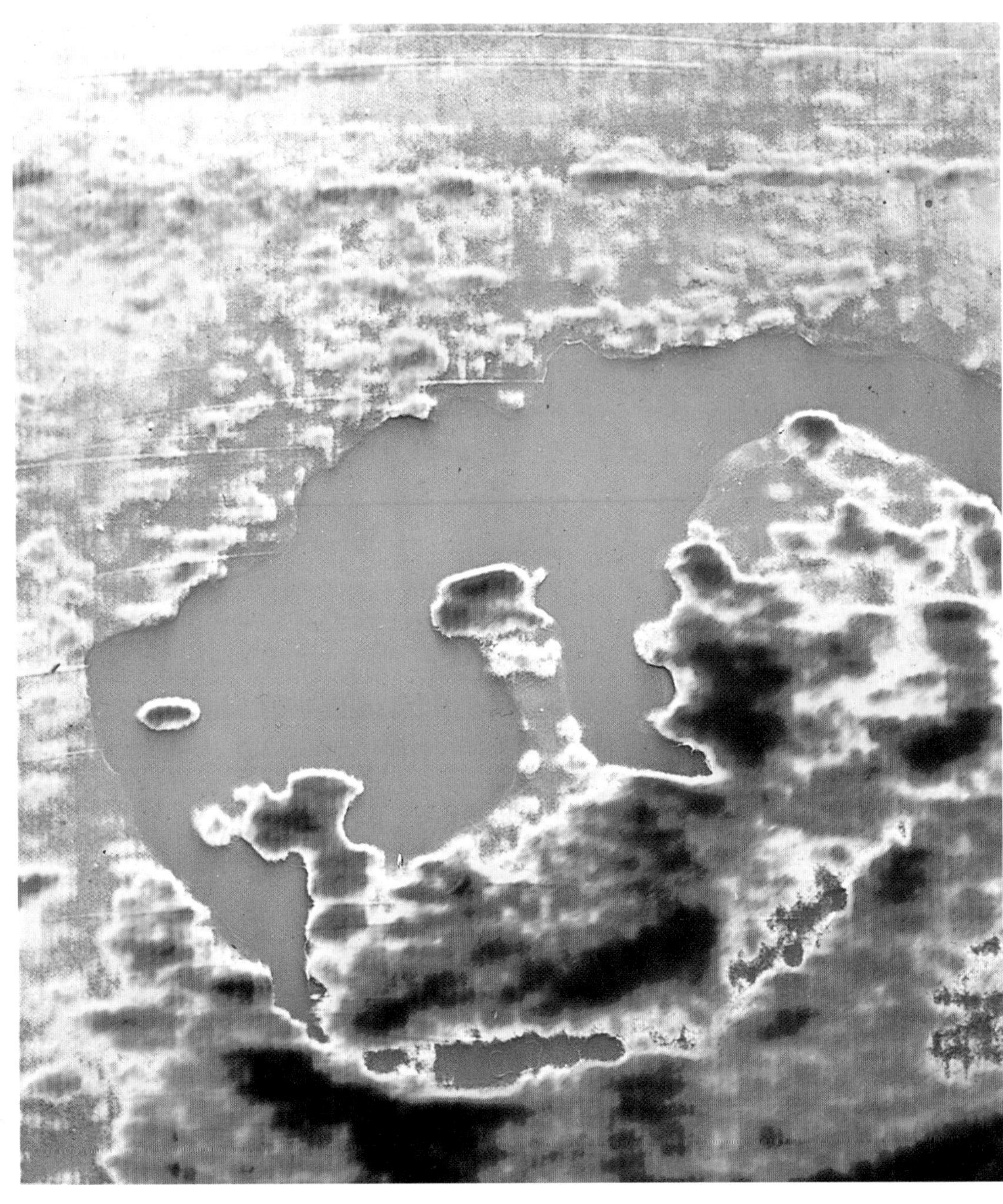

▲ *Child with Down's syndrome, a genetic disease resulting in distinctive facial characteristics and minor mental retardation. Older mothers have a greater risk of having Down's babies and may be offered prenatal diagnosis and counseling.*

◄ *An ultrasound picture of a 12-week human fetus in the womb. Ultrasound images allow doctors to take samples of fetal tissue early in a pregnancy to be used in genetic tests. At 12 weeks, the fetus is less than 10cm long and the mother is not aware of its movement. Therefore, if a genetic defect is found, an abortion will be less traumatic at this early stage.*

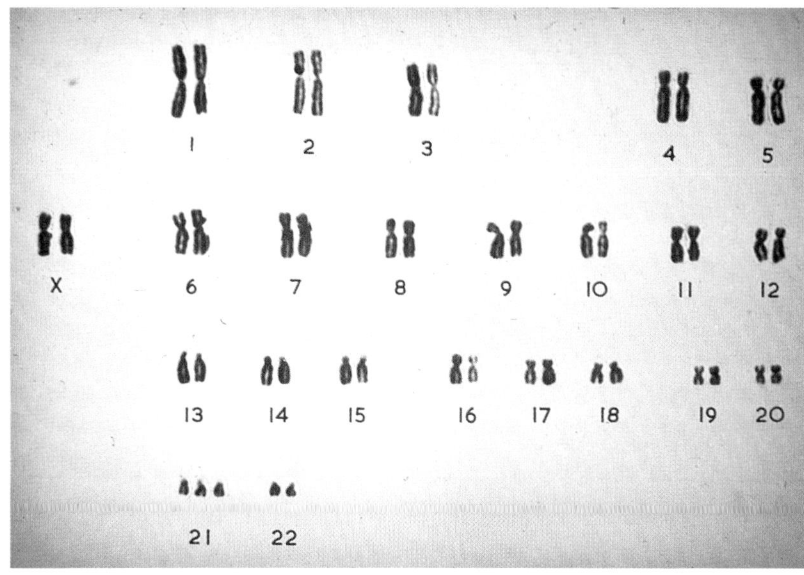

▲ *The pattern of human chromosomes in Down's syndrome revealed by a process called karyotyping. Doctors photograph test cells just before they divide, when individual chromosomes are separate and have distinctive shapes. Normally there are 23 pairs of chromosomes. The two X chromosomes in this karyotype show that the cells come from a female. The extra, third chromosome 21, present because of faulty chromosome separation shortly after fertilization, indicates Down's syndrome.*

Nowadays, after only 8–12 weeks of a pregnancy, doctors can use a technique called chorionic villus biopsy to take a sample of the chorionic membrane which surrounds the fetus. This is fetal, not maternal, tissue and so provides the same diagnostic information as the cells from amniotic fluid. Doctors can then perform an abortion much earlier, if required to do so.

When combined with genetic diagnosis, both amniocentesis and chorionic villus biopsy become much more useful. The DNA of the fetal cells can be examined for specific suspected defects. There are prenatal tests for Duchenne muscular dystrophy, Huntington's disease, cystic fibrosis, polycystic kidney disease, sickle cell anemia, hemophilia and an eye condition called retinoblastoma. Not all families at risk can be tested because of the need for a family pedigree of the disease. Those who can, however, are much better equipped to decide whether a pregnancy should be terminated. Many DNA probe tests can determine abnormality or otherwise with more than 99 percent certainty – a vast improvement on the 50–50 chance that parents used to face.

Genetic diagnosis might even be performed before a pregnancy is confirmed. With hormone stimulation, women will produce many eggs rather than one. When fertilized naturally following sexual intercourse, each will develop, after four or five days, into a blastocyst, a multicelled pre-embryo. These can be gently flushed from the uterus and one or two cells removed from each without causing any damage. Scientists could obtain enough genetic material from the embryo to conduct DNA probe assays. After the results were known, the unaffected blastocysts (stored in liquid nitrogen during the genetic analysis) could be thawed and replaced in the uterus. Scientists could analyze even earlier pre-embryos, containing only eight cells and obtained by in vitro fertilization (test-tube baby techniques). Such tests might be appropriate where there was a particularly high risk of disease, or strong religious or medical grounds for avoiding abortion.

Inner healing

Traditionally, medicines have been extracted from plants, animals or microorganisms (most of the antibiotics, for instance, are produced by microbes isolated from the soil). Biotechnology provides remedies that are to be found inside our own bodies.

Through genetic engineering, proteins responsible for maintaining health can now be produced in amounts large enough to study or use as drugs. Genetic engineers use recombinant DNA techniques to splice new DNA into a plasmid (see page ◀ 26) and then insert the plasmid into *E. coli*. The genetically engineered microbes grow to a high concentration (over 100 million organisms per cubic centimeter) in a fermentation vessel. Each organism produces hundreds of copies of the plasmid.

Only very small amounts of human proteins are usually needed to treat patients. One-tenth of a gram of insulin provides between 1,000 and 20,000 doses. The same amount of growth hormone represents daily treatment for about 10,000 patients.

The human proteins produced by genetically engineered organisms fall into two main classes. Firstly, there are those like insulin, growth hormone and Factor VIII, whose deficiencies cause diabetes, dwarfism and hemophilia, respectively. Treatment of these diseases aims to restore the natural levels of the missing proteins in the body. Traditionally, insulin has been obtained from the pancreas glands of cattle and pigs, growth hormone from the pituitary glands of human corpses and Factor VIII from donated human blood. Genetic engineering provides alternative sources – and sometimes substitutes – for these compounds that are often cheaper, more convenient and safer. Coronary thrombosis has for many years been treated with bacteria enzymes like streptokinase which destroy the blood clots blocking the blood vessels. Now genetic engineering has produced tissue plasminogen activator (tPA), a human protein which does a better job and with fewer side effects. tPA is one of a growing number of approved, genetically engineered drugs that have come onto the market.

There is a second class of proteins, the biological response modifiers, for which genetic engineering is the only realistic source. The biological response modifiers occur at very low levels in the body. In many cases, scientists have only recently become aware of their existence. Nevertheless, they are vital in controlling bodily processes.

► The growth market in human proteins. Treatment with recombinant human growth hormone caused Tracy to grow five inches in a year, pushing her closer to the normal height for her age. Growth hormone used to be extracted from pituitary glands of human corpses, until fears of contamination led to a ban on such a preparation. Soon afterwards, the US Food and Drugs Administration allowed doctors to use the genetically engineered version of the hormone. Human growth hormone may also accelerate the healing of wounds.

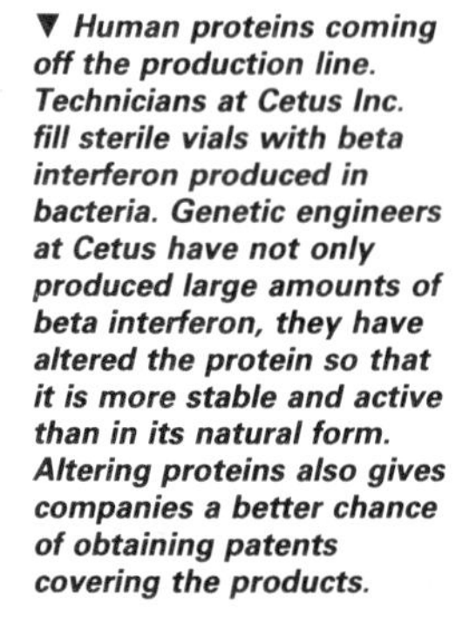

▼ Human proteins coming off the production line. Technicians at Cetus Inc. fill sterile vials with beta interferon produced in bacteria. Genetic engineers at Cetus have not only produced large amounts of beta interferon, they have altered the protein so that it is more stable and active than in its natural form. Altering proteins also gives companies a better chance of obtaining patents covering the products.

▼ Michael Sweeney, a diabetic who works at the American company, Eli Lilly, which produces human insulin from E. coli. Human insulin was the first genetically engineered medicine licenced for general use. Even though it is unclear whether human insulin has any real advantages to patients, it is replacing the insulin extracted from pigs and cattle which diabetics had used previously.

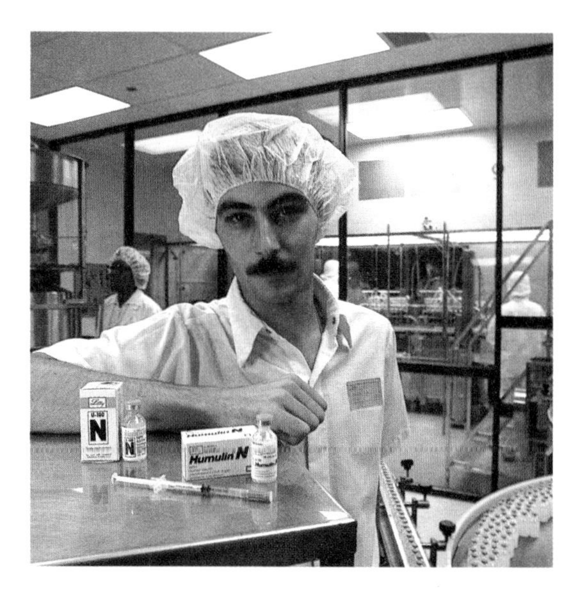

Alpha interferon, for instance, which is in the blood to guard against viral infection, is found in such low concentrations that it takes 90,000 donors to produce one gram of an inevitably highly impure preparation. Scientists have cloned the genes for proteins like interferon, interleukins (immune stimulants) and endorphins (natural pain killers) in microorganisms. Many types of these proteins have already undergone safety tests in animals and humans. Through genetic engineering, scientists now know that there are three main types of interferon – alpha, beta and gamma – and many subclasses.

Interferon is produced naturally by cells in the body that are infected by viruses. It diffuses into surrounding cells and protects them from infection. Many believed, therefore, that interferon would provide a cure for virus diseases or even cancer. Indeed, alpha interferon is effective against certain bone and throat tumors, hairy-cell leukemia and the AIDS-associated cancer, Kaposi's sarcoma. Unfortunately, it is less effective against the much more common cancers of the lung, breast or skin. It has been successfully used for treating herpes and hepatitis, and, taken as a nasal spray beforehand, it can stop common-cold infections. However, interferon is also responsible for all the distressing symptoms of a cold. Gamma interferon is being tested for effectiveness against rheumatoid arthritis, systemic lupus erythematosus and multiple sclerosis. Using genetic engineering, scientists have also investigated the functions of interleukins (also known as cytokines or lymphokines), which stimulate the immune reaction, as well as neuropeptides – small, natural protein painkillers – and many growth-controlling human proteins.

◀ Insulin is a small protein made up of two amino acid chains which are held together by disulfide bonds (yellow). Scientists at the Novo company in Denmark have produced a single-chain insulin which is active and seems to be absorbed more easily in the gut than is the normal protein. This raises the possibility that diabetics could take insulin orally with their meals, rather than by injection.

Transplants and artificial organs

An artificial pancreas would be useful in the treatment of diabetes. The pancreas regulates sugar uptake in the blood by altering the balance of insulin and glucagon, the two hormones it produces, in response to levels of glucose in the blood.

One prototype artificial pancreas links the detection of blood glucose directly to the supply of insulin, using both biotechnology and electronics. The usual test for blood glucose is an enzyme-based diagnostic aid. A positive test is indicated by a dye changing color when electrons are transferred to it from glucose. If, instead, the electrons are transferred to an electrode, an electrical signal is produced when the enzyme encounters glucose. A growing number of diabetics measure their blood glucose using miniature glucose biosensors based on this principle. One version looks like a ballpoint pen, but with the nib replaced by a sharp, disposable point for puncturing the skin. The body of the "pen" holds the electronics and a liquid crystal digital display panel.

To produce the artificial pancreas, scientists plan to link a fine wire biosensor to an electronically activated insulin pump. A fine tube from the pump takes the insulin into the bloodstream. The pump would be activated by an electronic signal from the glucose biosensor. Insulin added to the blood would stimulate the uptake of glucose by cells and hence reduce its concentration in the blood. The sensor, detecting this change, would then switch off the pump. Researchers hope that "closing the loop" in this way will ultimately lead to better control of diabetes.

The best source of cells for transplantation is perhaps the patient's own body because the immune system does not react against it. Skin cells from the legs or back, for instance, are often used to replace tissue damaged by burns or scalds. Cell grafting is also useful in treating certain types of leukemia, in which malignant white blood cells outgrow the normal ones. The usual treatment is called bone marrow rescue. Doctors kill or remove diseased cells and replace them with healthy ones. A sample of the patient's bone marrow, an important source of blood cells, is removed and frozen. After doctors have administered high doses of irradiation or chemotherapy to kill all the tumor cells in the patient's body, the stored marrow is thawed and infused back into the bone. If all goes well, only healthy cells grow from the marrow. However, if there are living tumor cells in the stored marrow this may lead to a recurrence of the cancer.

To prevent this happening, researchers treat the stored marrow with immunotoxins (toxins linked to antibodies), which kill the tumor cells. In experiments on mice, only 10 percent of normal cells were damaged by the treatment, leaving more than enough to reestablish a healthy blood cell population when the marrow was reinserted into the bone. Similar techniques might be used for any blood condition in which diseased and healthy cells can be distinguished.

Related methods have been used against cancer. A genetically engineered protein, interleukin-2, acts on transfused blood cells to stimulate the patient's immune response to the disease, converting normal immune cells into "killer" cells. The patient's blood cells are grown for several days in the laboratory with interleukin-2 until they are aggressive enough to kill tumor cells. They are then replaced in the blood. Interleukin-2 is administered to maintain the aggression. This treatment, called adoptive immunotherapy, can reduce even established tumors, but makes the patient feel very ill.

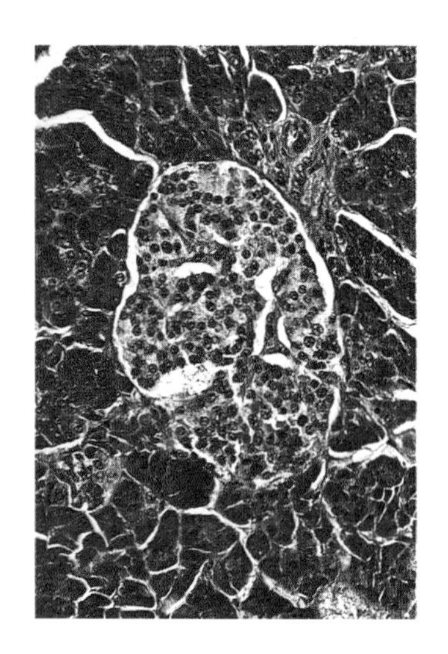

▲ Hormone-producing cells of the pancreas. In diabetes, these cells are defective. Scientists have encased pig pancreas cells in an inert coating and implanted them into a human pancreas to help restore its function. Implants of brain cells from human fetuses could be used in a similar way to treat Parkinson's disease.

▼▶ Surgeons collecting bone marrow from donors to treat patients who have immune system defects. Since the bone marrow cells are immune cells, they might reject the patient (usually patients reject grafts, not the converse). To ensure that treatment is successful, the marrow is pretreated with a panel of monoclonal antibodies (from the vials) to knock out those cells which react against the patient.

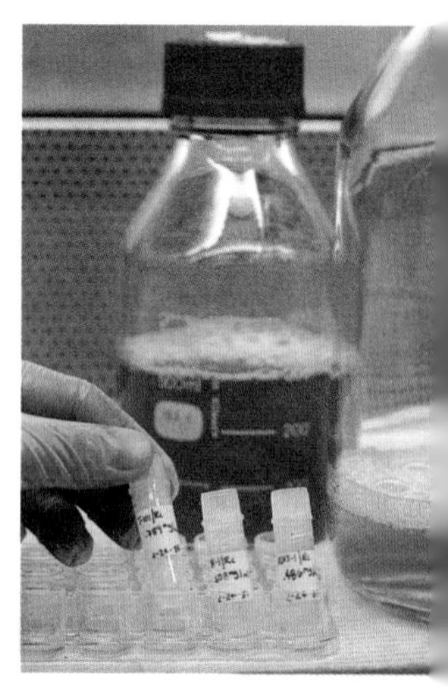

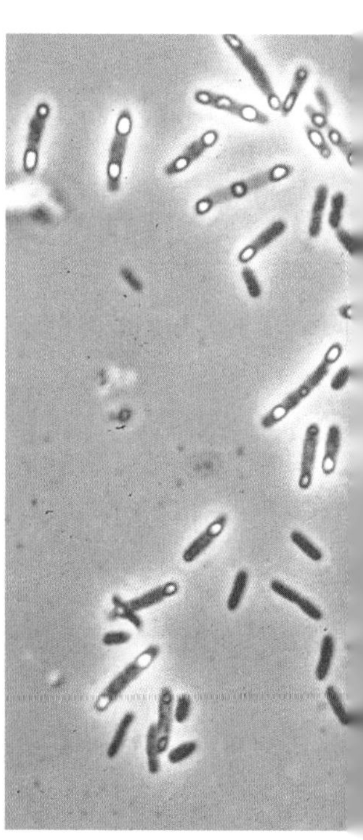

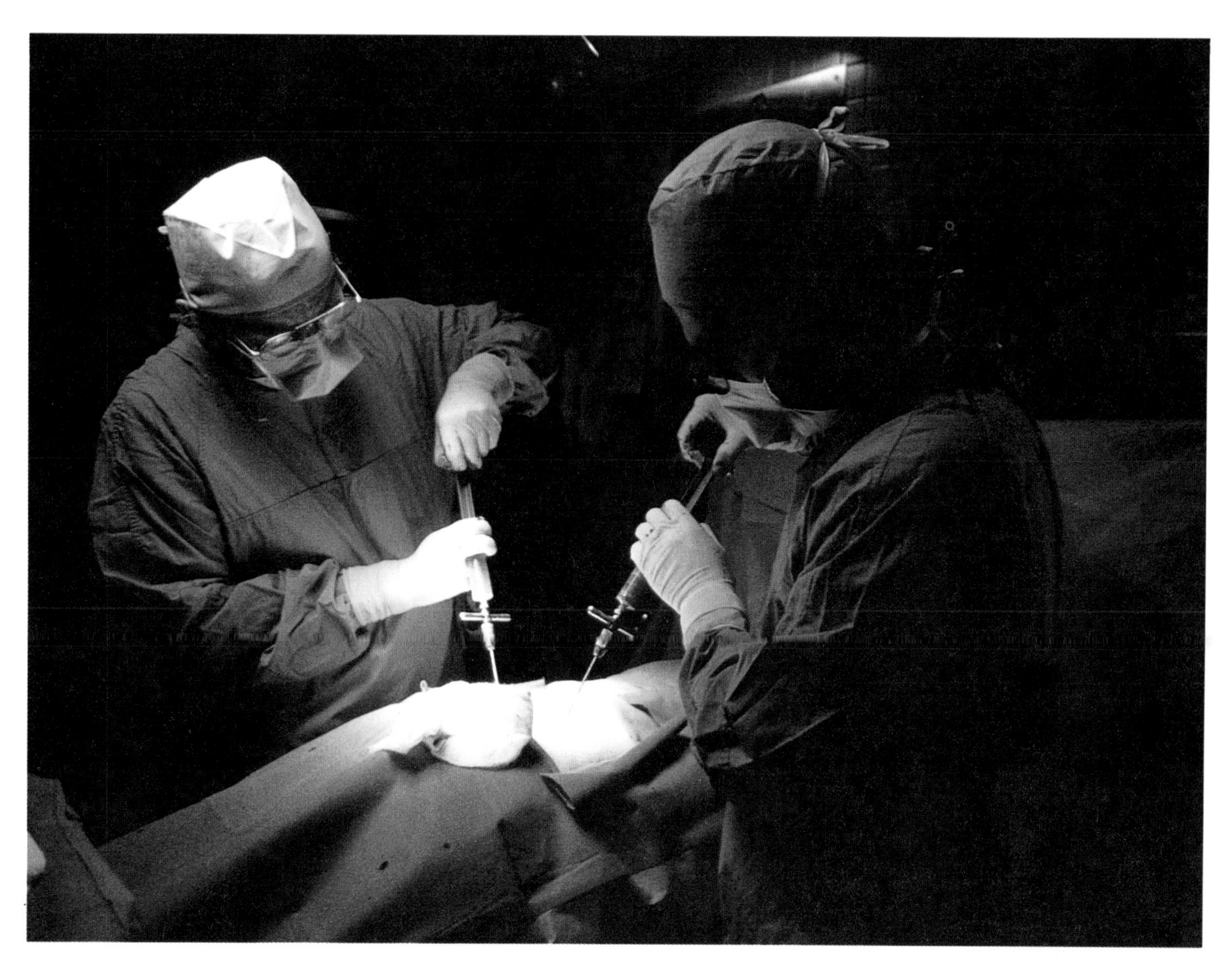

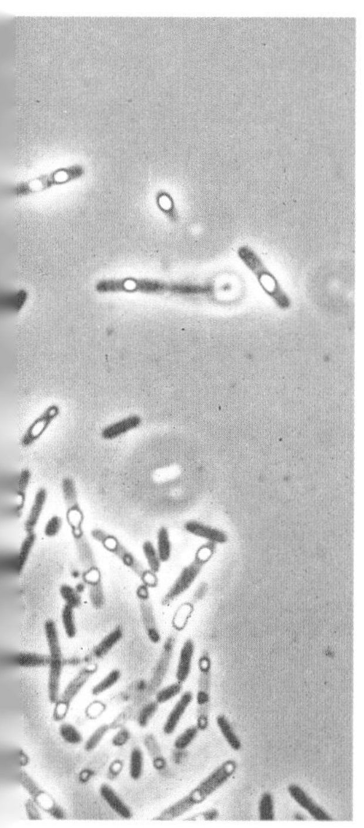

Human protein side effects

Interferons and other human proteins produced in genetically engineered organisms have toxic effects when used as drugs. At first, impurities were thought to be the cause. Better purification methods helped somewhat. Other side-effects were accounted for by differences between proteins produced in bacteria and the human versions. For instance, some proteins have extra sugar molecules added when they are processed naturally. Bacteria lack the biochemical pathways for doing this. So yeast and mammalian cells which can also be grown in culture were used to produce more realistic proteins. Even then, harmful effects remained. Toxicity seems to be an intrinsic property of some molecules.

Scientists are starting to be able to understand why it has been so difficult to use these rare proteins therapeutically. The normal response of the body to virus infections or cancer involves a complex network of protein factors. Each of them has many different activities in the body, some of which are responsible for the side-effects. Some scientists are now using "cocktails" of two or more protein drugs in clinical trials. Others are using knowledge of the proteins' structure to work out which regions are associated with desirable activities and which with side-effects. This research is aimed at designing new, less toxic compounds that can be manufactured by genetic engineering.

◄ Genetically modified E. coli cells produce and accumulate interleukin-2, a rare human protein which regulates the immune system. During the growth of E. coli, so much protein is produced that it forms crystals in the cells. There are at least six different interleukins controlling separate aspects of the immune system. Since scientists have only recently obtained usable amounts of these proteins, they do not completely understand how they work. Even when the proteins are almost 100% pure, they cause unwanted side effects in trial patients..

Ultimately, genetic engineering may be applied to the DNA of human cells to treat inherited disease

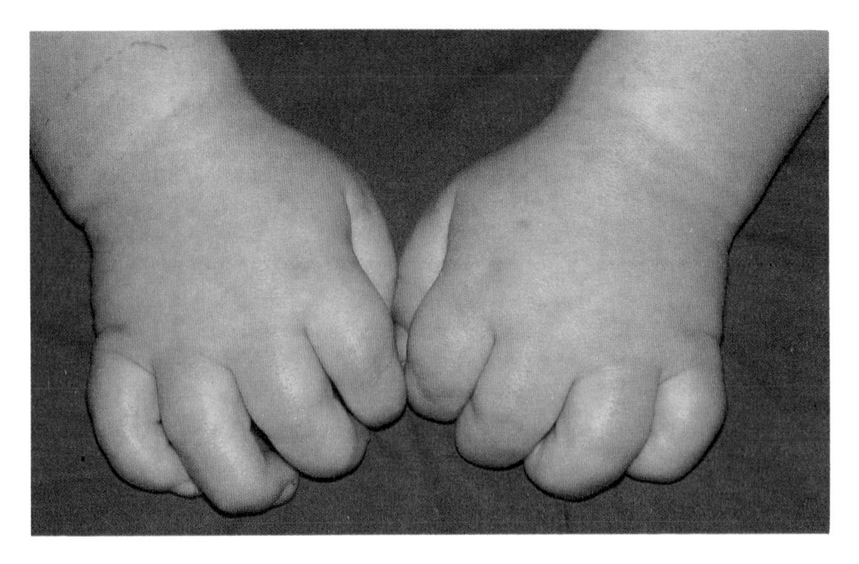

◀ Swollen claw-like hands are one of the symptoms of Hurler's syndrome, a form of gargoylism, and another candidate for gene therapy. There is no method of treatment to prevent sufferers becoming severely mentally retarded and dying before the age of 20. Even for such severe ailments gene therapy is only feasible if the disease is known to be due to defects in single genes and if no other cure exists.

▼ Liver cells from a patients with Gaucher's disease, a severe genetic disease in which fatty deposits accumulate in cells because of a defective enzyme. Gene therapy may be a way to treat it. Other candidate diseases for gene therapy includes Lesch-Nyhan syndrome, in which self-mutilation is one of the most distressing symptoms, and Tay-Sach's disease, which is prevalent in certain Jewish communities.

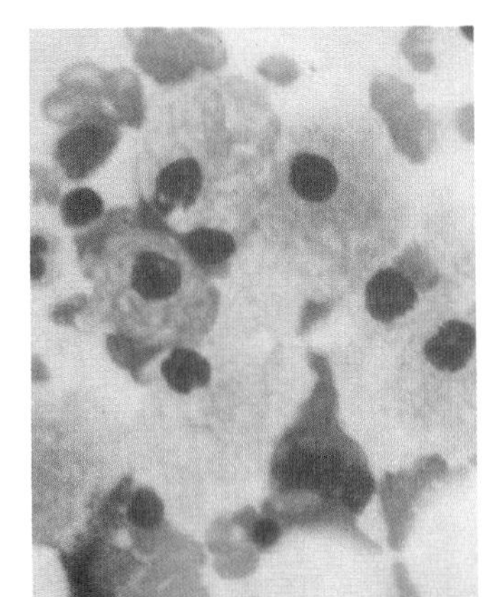

Gene therapy

A patient's cells can be taken from blood or bone marrow, treated in various ways and then replaced. Ultimately, genetic engineering may be applied to the DNA of the cells themselves in order to treat inherited diseases. This technique is called somatic cell gene therapy. Somatic cells are non-reproductive cells such as those of our skin, brain, bone marrow and blood. Any genetic changes produced in these cells will not be passed on to our children. Only diseases caused by mutation in single genes can be treated by somatic cell gene therapy.

Researchers into the method have used vectors – collectors and carriers of genes. An important class of vectors are retroviruses. Their genetic material, RNA, is copied into DNA when they invade animals and human cells. To produce effective vectors, scientists splice out essential genes from viruses. This both renders the virus harmless (unable to infect cells unless assisted) and makes room for other genes to be added. Using the test-tube techniques of genetic engineering, scientists can produce a few vectors in which normal human genes are inserted.

For a successful transfer of genes, hundreds of millions of copies are needed. Therefore, the vector must multiply. This is done with the assistance of a "helper" virus with its genes intact. Cultured cells are infected with both viruses at the same time and DNA copies of their RNA are made. These enter the nucleus of the cell and take over the biochemical machinery for producing RNA. The "helper" virus RNA is exported from the nucleus and packaged automatically in the coat protein. Each infected cell produces hundreds of copies of the vector RNA, each wrapped in protein. The process can be repeated in neighboring cells of the culture.

When sufficient vectors have been produced and packaged, they are purified ready for use in somatic cell gene therapy. Again, the virus infects cells (blood and bone marrow cells in the experiments performed in animals) and produces a DNA copy which enters the nucleus. Since the "helper" virus is not added, the vector cannot multiply. Instead, all or part of it may be integrated into the DNA of the cells, and thus add a copy of the unaffected gene. If this happens, any daughter cells will also inherit the vector.

So far, the method has been successful in some experiments on animals. One of the first genetic diseases that may be treated by somatic cell gene therapy is severe immune deficiency. In the absence of an enzyme called adenosine deaminase (ADA), many cells of the immune system do not function, leaving the patient vulnerable to all sorts of infections. The gene coding for ADA has been transferred into monkey bone marrow, and the enzyme produced, albeit at a low level, when the cells are returned to the animal. Scientists have also managed to transfer and express genes for antibiotic resistance in human cells in culture. (This was only an experimental system and the cells were not put back into human beings.)

Germline cell therapy

One of the difficulties with somatic cell gene therapy is that people who undergo the treatment still risk passing on the defect to their children. If gene therapy was extended to reproductive, or germline, cells, this might enable genetic defects to be cured and not inherited. The technique has been shown to work in animals. DNA injected into the nuclei of single-celled embryos of cows, mice and sheep is sometimes integrated into the chromosomes. These new genes can direct the production of new proteins in the animals which develop from these embryos. The animals are otherwise, apparently normal.

Most of the technology for making similar additions to human embryos has already been developed. Genes from various sources can be inserted, via retroviral vectors, into human cells. Human eggs and sperm are relatively easy to obtain and can both be stored by deep-freezing. Single-cell embryos are handled regularly during in vitro fertilization (test-tube baby techniques). They can be grown in culture for several days, stored for later use or implanted into a woman's uterus. However, gene manipulation techniques are not yet well enough understood for germline cell therapy to be tried on human subjects.

Biotechnology and Society

6

The public's view of biotechnology...Controlling experiments...Gene therapy: Pros and cons...The commercial connection...Biology as property...The new agricultural revolution: Its effects in industrialized and developing countries...Personal genetic profiles: Who has a right to know?

Many biotechnologists would say that biotechnology is not a science at all, but a collection of practical techniques, like genetic engineering and fermentation, aimed at solving real problems. It can feed the hungry, make short people taller and, one day perhaps, provide remedies for presently incurable diseases. As with any technological advance, biotechnology will change the way human beings live their lives and think about the world.

In the early 1970s genetic engineering was an experimental technique. It could only be performed in one very specialized laboratory organism (*E. coli*). The work was tedious and difficult to perfect. Its first product, insulin, would not be sold for another decade. Yet almost as soon as the results of the first scientific investigations had been published, there was a rush of public concern. Although scientists were only experimenting with bacteria, many people felt that they were compromising the sanctity of life. Altering genes in microorganisms was seen as the prelude to parallel experiments on animals and perhaps on humans. If genetic engineers could alter genes, the argument ran, they could influence human characteristics, both physical and mental – they could change human needs and desires, our bodies and souls, and alter human society.

The public perception

Genetic engineering has been making headlines since 1972 when scientists working at Stanford University in California first spliced DNA molecules. Perhaps more than any other scientific development in its earliest stages, this breakthrough aroused huge public interest.

"Genetic engineering" is scientific shorthand for "methods for joining pieces of DNA in the test tube". An alternative phrase, "in vitro recombinant DNA techniques" is more accurate and less open to misinterpretation. Unfortunately, the general public knew little about DNA or how it could be recombined in vitro. So "genetic engineering" stuck as the phrase used to describe the new technology. "Engineering" has connotations of preconceived design and practical construction, notions that to many people are incompatible with the natural processes of reproduction, heredity and growth, which are the substance of "genetics". Another term "cloning", especially in the phrase "cloning human genes", caused further communication problems between scientists and the public, raising visions of Huxley's Brave New World *or the spector of eugenics.*

▼ Francis Crick (left) and James Watson who elucidated the structure of genetic material. Watson holds a copy of their historic scientific paper which first described the double helix of DNA. Behind them are many of their colleagues who built upon their work to broaden the scientific basis of biotechnology.

▼ A pile of biotechnology books forming part of the reference collection of the US Library of Congress. While the politicians may not read them, their researchers and advisors certainly will. Biotechnology may concern business people, farmers, industrialists, lawyers, clergy and many others besides in non-scientific professions.

While 42 percent of a sample of people morally disapproved of genetic engineering in human beings, an even greater number approved of its use for improving intelligence or for cosmetic purposes

Essentially, people wanted to know, "Is genetic engineering dangerous?" Scientists too were concerned that safety aspects might not have been fully considered. While one group of researchers had plans to try to insert a monkey virus called SV40 into *E. coli*, others felt this was risky, because SV40 sometimes causes cancer in mice and rats, and also because *E. coli* is a natural inhabitant of human intestines. In an attempt to alleviate public anxiety and preempt real or imagined disasters in recombinant DNA experiments, a group of American scientists appealed to the President of the US National Academy of Sciences to set up a committee of investigation. In 1974 the committee proposed that almost all research involving recombinant DNA should be suspended until the possible dangers had been asessed.

An experimental ban

A letter announcing the moratorium was published in 1974 in the US journal *Science*. Later that year, the US National Institutes of Health (NIH), which funded the majority of the research, began to draw up a set of guidelines to govern genetic engineering experiments. With the publication of the guidelines in 1976, the moratorium ended. Experiments were categorized according to the magnitude of potential hazards. Thus, work involving the DNA of lethal microorganisms like the anthrax bacillus and smallpox virus necessitated the most stringent precautions, whereas some experiments only required adherence to normal laboratory practise. The guidelines proposed that genetically engineered organisms should be contained both physically and biologically. Physical containment included such things as airlocks on laboratories and air filtration, and was intended to prevent genetically engineered organisms from escaping. Biological containment entailed using only "crippled" organisms for research, so that if they did escape they could not grow or pass on their genes.

The guidelines applied only to US government-sponsored research, but it was understood that industry, too, would restrict its experiments accordingly. Other countries adopted similar codes of practise or imposed even tighter controls. In the Netherlands, there was virtually a total ban on genetic experiments, forcing companies affected by the ban to establish research centers abroad.

By the end of the decade a combination of pressure from scientists who increasingly resented outside interference with their research and pressure from industry and industry-related government bodies resulted worldwide in major revisions to the guidelines. Experience, it was argued, had shown that genetic engineering confined to the laboratory was, on the whole, safe. For many uses, as vaccines, soil inoculants or crops, however, organisms need to operate outside the artificial confines of a laboratory.

Ecologists feared that engineering human viruses or organisms causing plant diseases, for instance, could create strains giving rise to other diseases that might be more difficult to treat. Genetically engineered plants might also be able to transfer their new genes to closely related weeds. Similarly, natural mechanisms of gene transfer among bacteria in the soil might enable the spread of newly introduced genes. In 1985, the US Environmental Protection Agency (EPA), which had long deliberated over whether genetically engineered organisms should be used in agriculture, allowed the Mycogen Company from California to use killed, genetically-engineered organisms as pesticides. In 1986, a British experiment involving harmless viruses tested the first use of live organisms in an open environment.

▼ University of California researcher, Steven Lindow, standing in the field where he had hoped to show that "Ice-minus" bacteria could protect strawberry plants from frost damage (⧫ page 209). He had to wait from 1984 until 1987 to perform the field trials. The delays were caused by legal proceedings and by local opposition to the tests. In one incident, a young mother approached Lindow from a crowd of objectors. She held her child up to him and asked "Why do you want to kill my baby?" Most scientists are not used to such emotional public response.

▲ Jeremy Rifkin representing the environmentalist views of the Foundation for Economic Trends to a gathering of scientists in Florence, Italy. Rifkin's persuasiveness won much public support for his anti-biotechnology stance. Most scientists, however, remained unconvinced, but found it difficult to explain their side of the argument.

► Geneticists check tobacco on a test site in Wisconsin, USA, as part of the first trial of genetically engineered plants. The public is much less worried about the recombinant DNA crops than about microorganisms. This may be because the advantages of genetically engineered crops (although perhaps not those of tobacco) are more tangible than the benefits of better microbes.

The DNA controversy

As genetic engineering methods improved, viruses, yeasts, fungi, plants and even animals could be manipulated, and more applications of those organisms were conceived. So far, none of the supposed horrors of genetic techniques have been realized even after hundreds of thousands of experiments.

The confidence of the biotechnology professionals is not wholly shared, however, by the general public. In a recent American survey, 52 percent thought that genetically engineered products probably did represent a serious danger to people or the environment. Yet, paradoxically, 66 percent believed that genetic engineering would improve their quality of life.

People's attitudes to biotechnology tend to vary according to how they are asked to think about the question. Questioned generally about the morality of genetic therapy, 42 percent felt that changing the genetic makeup of human cells was morally wrong, but twice as many approved the use of gene therapy to save children from inheriting fatal genetic diseases, and more than three quarters approved its use to prevent nonfatal genetic diseases. Surprisingly, as many as 44 percent said they would approve of genetic engineering being used to improve levels of intelligence or even physical appearance. Seventy-eight percent said they would be willing to undergo genetic therapy to prevent serious or fatal genetic disease.

Gene therapy – pros and cons

One of the fears of scientists working in somatic cell gene therapy (performed on blood cells; ◆ page 110) is that some cells harbor viruses that might act as "helpers" and enable the vector to spread. Another worry is that the insertion of new DNA sequences into chromosomes might activate oncogenes – latent genes associated with cancer. There is no scientific evidence that this sort of damage occurs – the engineered human somatic cells appear normal. Some researchers expect that by 1990, somatic gene therapy will have entered clinical trials.

For germline cell gene engineering (performed on the reproductive cells; ◆ page 110), neither the risks nor the benefits are clear. Before the treatment could be considered, scientists would need to reduce the risk of a misplaced piece of DNA triggering cancer genes. The chances of interfering with an essential gene function are much higher in germline than in somatic cell therapy. Damage to a gene coding for an essential brain function, for instance, would not affect the viability of bone marrow or blood cells, but it would certainly affect the embryo.

There is also a risk that germline techniques could be used for eugenic purposes – to breed human beings selectively. Most parents

▼ Making a withdrawal from a specialist sperm bank established by feminist groups. One way or another, we all choose our sexual partners and thereby participate in selective breeding. If people understood more about genes, would they screen the DNA of prospective partners for desirable traits?

want their offspring to be "normal", by which they generally mean without physical or mental deformities. It may be possible in the near future (when more is known of the human genetic sequence) to tell from a person's genes exactly what characteristics they will inherit. If genetic therapy is by then sufficiently advanced, scientists may be able to alter characteristics and correct "defects" (however defined).

For some inherited diseases, genetic therapy would be beneficial because of the cost of the untreated condition to patients, their families and society. But there are very few conditions where germline therapy would be preferable to somatic cell treatments. Furthermore, as scientists learn more about our genes, treatments other than gene manipulation may be developed.

◀ Stanley Cohen and his colleague at Stanford University, Herbert Boyer, carried out the first genetic engineering experiments. Later, they joined forces with investors to set up Genentech Inc., a company that would exploit the technique. Genentech now sells millions of dollars' worth of protein drugs a week.

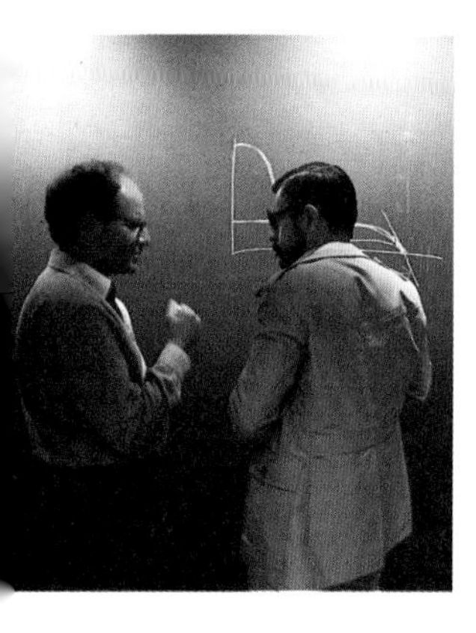

◀ Wally Gilbert (left), Harvard professor and inventor of one of the techniques for sequencing DNA, became scientific director of a biotechnology firm called Biogen. After several years with Biogen, he returned to Harvard and later established another company, Genome Corp., for the express purpose of obtaining the entire sequence of human DNA. Unfortunately, the new company failed to attract sufficient investment and closed in September 1988.

◀ Human gene therapy will be possible, sooner or later. The question is, how to make best use of it? Our present knowledge does not even allow us to decide which genes are "bad". Sickle-cell anemia has spread because its genes are beneficial in malaria-stricken areas. Cystic fibrosis genes, too, might be beneficial, explaining why the mutation is so widespread. In some societies where male children are preferred, the absence of a Y chromosome might even be considered an example of a genetic defect.

The commercial connection

Not only did the development of genetic engineering focus public attention on biology for the first time, it also revolutionized the social and commercial role of the biologist. Those who could clone genes suddenly found themselves in great demand as consultants to large drug and agrochemical companies. Others were employed to establish new research units within firms. Many scientists left poorly paid posts in underequipped university laboratories to take up more lucrative and better supported positions in industry.

As more and more of the leading researchers followed this route in the late 1970s and 1980s, government agencies responsible for funding science began to be fearful of specialized scientific knowledge becoming the property of commercial monopolies whose primary motivation was profit and not the public good. Research within these firms would tend to be "applied" – directed towards a particular end – rather than pure research. Findings were kept secret, or publication would be delayed, to retain a commercial edge over competitors. Several European countries were worried that the high salaries paid by firms abroad would encourage top scientists to vote with their feet and lead to a biotechnological "brain drain".

In the United States growing public awareness of genetic engineering and its commercial potential was soon reflected in the stockmarkets. On 14 October 1980, with a product folio consisting largely of promises, the first genetic engineering company, Genentech, issued shares for sale to the public. Thousands of investors scrambled to buy, encouraged by press reports of the companys' scientific breakthroughs. In the first 20 minutes of trading, the price of Genentech shares rose from $35 to $89, a stockmarket record. Suddenly, Genentech, still a relatively small concern and with no real products, acquired a paper value of over $200 million and had raised vast sums of money for its research. Today it is the largest biotechnological company in the world and is diversifying into pharmaceuticals.

Other biotechnology companies followed Genentech to the stockmarket. In the United States, Cetus, Molecular Genetics, Genex, Hybritech and Biogen all "went public" between 1981 and 1983. Though less spectacular than in the case of Genentech, the sales of their shares hugely financed their research projects. Soon, breakthroughs in biotechnology featured on the financial pages of daily newspapers as often as in the science and medical columns, impinging more and more on people's lives and providing detractors with further ammunition. The mystery that surrounds much of science was compounded by suspicions that, not only did genetic engineers want to tamper with life itself, but they were driven to it by greed.

Biology as property

The compounds produced by genetically engineered organisms are very highly valued. So, too, are the organisms that produced them. Not surprisingly, companies were keen to prevent their rivals from using their highly productive strains. In other fields of high technology endeavor, patents are used to prevent competitors from copying. A patent guarantees an inventor a limited period of exclusive use of the invention provided that a complete description of it is lodged with the patent-issuing office. Biological organisms had been specifically excluded from patent coverage because they were living and "products of nature".

Other complex questions concern the use of tissues from human embryos. By advanced surgical techniques clinicians can extract certain tissues from tiny human fetuses and implant them into adults to restore pancreas or liver function. Brain cells have also been transferred from fetuses and implanted in the brains of sufferers from Parkinson's disease. The advantage of fetal tissue is that, unlike other foreign tissue, it is not rejected by the body's immune system. In effect, the availability of these techniques creates a market for fetal tissue. People dying of diseases caused by atrophy of an internal organ and desperate to find a tissue donor would undoubtedly be prepared to pay a high price. There are fears that fetuses might be deliberately aborted in order to produce the valuable life-saving tissue.

▼ Workers shipping tissue plasminogen activator (tPA), a genetically engineered treatment for coronary thrombosis, to hospitals. When Genentech first sold tPA in 1987, it cost over $2,000 a shot. The firm claimed that tPA had fewer side effects than cheaper drugs and that patients should spend less time in hospital.

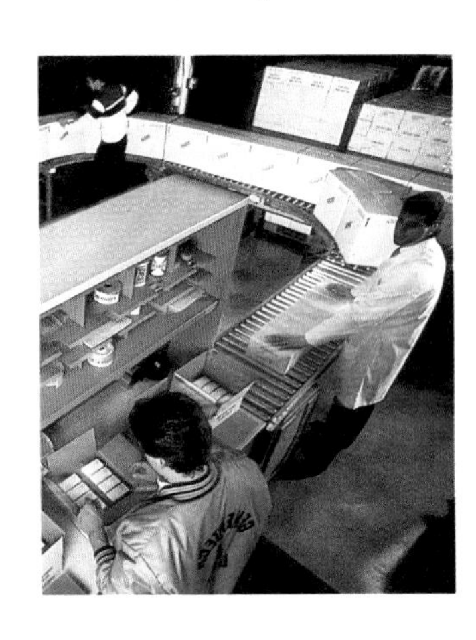

Further problems of ownership surround the Human Genome Project, the plan to map and find the sequence of the three billion bases of human DNA. The sequence information will undoubtedly be of great medical importance, helping scientists and doctors to understand the genetic basis of growth, development, normality and abnormality. However, many investigators are worried that the involvement of industrial research groups will mean that the information will not be freely available. Genome Corporation, a company established specifically for the project, but which later went out of business, had announced its intention to use the copyright laws to protect and extend its commercial interests. It aimed to treat the DNA that it sequenced as a string of letters which an interested party might buy but not sell. The company had estimated that it would still be cheaper for others to obtain the sequence information in this way than to acquire it by themselves. The practice of deliberately withholding data runs counter to the academic principles on which science is traditionally based.

▲ Ananda Chakrabarty outside the US Supreme Court holds the patent certificate that made legal history and propelled biotechnology along the path of commercialism. The flask in front of Chakrabarty contains his invention, a culture of genetically altered bacteria which "eat" oil spills. Many other microorganisms have since been patented. Genetically engineered plants and animals are also now patentable following successful legal challenges in the US courts.

► The first patent on an animal was granted to Harvard University for a genetically engineered mouse produced by faculty member Dr Philip Leder. Lawyers are divided as to whether patents on biological products will withstand future challenges in the courts.

Exploiting the sick

Firms making biological drugs have been attacked for taking advantage of the sick. When the gene for interferon was cloned in 1980, interferon became available for the first time in usable amounts. Rumors at the time suggested that there was a thriving blackmarket in the protein, with rich buyers, desperate in their attempts to treat cancer, paying huge sums to obtain what was still a very experimental drug. In 1988, the cost of a 10-year course of treatment with human growth hormone produced by recombinant DNA technology for a child suffering from dwarfism was between $10,000 and $15,000 a year. Human growth hormone extracted from the thymus glands of corpses, the earlier method of treatment, cost only a little more to purchase but was much dearer to produce. Drug companies justify their prices on the basis that new treatments are safer and are expensive, in both time and money, to develop.

A patent for life

Ananda Chakrabarty, a young researcher with the General Electric Company in the USA challenged the laws which said that life could not be patented. Chakrabarty had developed a strain of Pseudomonas bacterium to clear up oil slicks. The bacterium was genetically altered, but without using recombinant DNA techniques. Chakrabarty had discovered a process by which four plasmids from different sources could all be inserted into a single bacterium so that it could degrade four components of oil. This process could be patented but Chakrabarty also wanted to protect its product, the bacterium. His original patent application was filed in 1972, but it was not until 1980, and after several appeals, that a judgement was finally delivered. The US Court of Customs and Patent Appeal ruled, by a majority of five votes to four, that Chakrabarty's Pseudomonas strain was patentable. It did not matter, they said, whether something was living or not, so long as it was "the result of human ingenuity and research".

Some philosophers have wondered whether the Chakrabarty decision might not have raised bigger questions than it answered. If an organism can be owned because it is genetically engineered, what about a genetically altered or engineered human being? Parts of the human body have already been patented and are being sold. There is a growing trade in human tissue. A Los Angeles man, John Moore, was diagnosed as having a rare disease called hairy cell leukemia. After a successful operation to remove Moore's spleen, his doctor cultivated the tissue from the organ artificially. He found that Moore's cells produced a variety of interesting and potentially valuable proteins and in 1979 he applied for a patent on a type of cell called "Mo" which he had isolated from the spleen tumor. Subsequently, the cells were licensed to commercial firms, who paid the doctor handsomely. In 1984 Moore sued the doctor claiming that the cells had been misappropriated. The claim was rejected but Moore is appealing against the decision.

This case prompts all kinds of questions. Who owns human cells and tissues? Can we sell tissues and organs? Do surgeons who remove parts of our bodies have the right to dispose of them as they please? What of children's cells and tissues? Do parents have any rights over these?

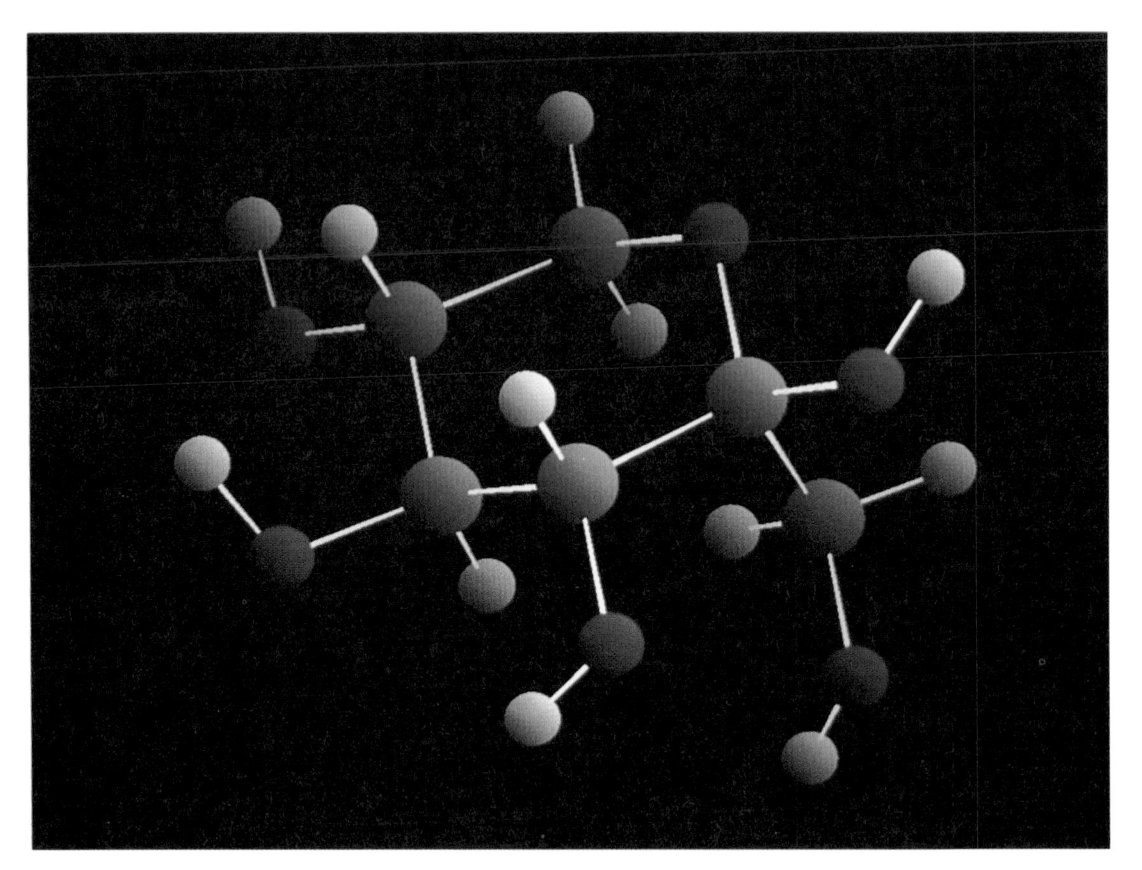

▶ *A molecule of fructose, the main component of the sweetener HFCS. Biotechnology companies have developed large-scale processes involving enzymes to produce hundreds of tonnes of fructose from maize. Fructose is cheap and has replaced sugar as the sweetness ingredient of many food products.*

Rich and poor

In the late 1960s an artificial sweetener, high fructose corn syrup (HFCS) was produced, one of the first biotechnological substitute foodstuffs. HFCS is made using two sorts of enzyme. Glucoamylase breaks down starch from corn (maize) into glucose, and glucose isomerase converts glucose into another sugar, fructose. Glucose is less sweet than sucrose (table sugar) but fructose is one and a half times sweeter. The conversion of corn starch into fructose greatly increased the sweetness of the product and HFCS became widely used, particularly in soft drinks. Aspartame is another biotechnologically-produced sweetener. Under the tradename Nutrasweet, it is a component of soft drinks, diet products and many other foods. Aspartame is a combination of two enzymatically-linked amino acids. It is nearly 200 times sweeter than sucrose and has virtually no calories. The sweetest edible substance known (2,000 times sweeter than sucrose) is a protein called thaumatin produced by the fruit of a West African plant. Genetic engineers have cloned the gene for this protein into microorganisms and can now produce it in large amounts.

A US government report estimated that HFCS production saved the country $1.3 billion in sugar imports in 1980. The impact of aspartame has been similar. But reduced demand has devastated the sugar market. The price of raw sugar has fallen disastrously, threatening the livelihoods of millions of cane farmers in the developing world. In the Philippines in 1980, export revenue from sugar fell from over $600 million to under $250 million and 500,000 workers had to be relocated or redeployed to produce other crops.

To make matters worse for them, biotechnology can also increase production of alternative crops, such as rice, that sugar-producing countries have been forced to turn to. By inoculating fields with blue-green algae, much more rice can be grown in the same area of paddy. Consequently, traditional importers of rice such as India or Indonesia are either becoming self-sufficient or even net exporters of the crop, while countries like Thailand and Pakistan, who used to be net exporters of rice, have lost their traditional markets. Vast amounts of rice have had to be stored or destroyed and at the same time the market price has fallen dramatically.

▼ *An abandoned sugar-cane processing factory in Barbados, West Indies. The biotechnological production of sweeteners such as HFCS, aspartame and thaumatin has had a far-reaching impact on the world's sugar trade. By reducing imports and using up surplus home-grown stocks of maize, biotechnology has benefited the richer countries like the USA. But there have been massive revenue losses for states such as those of the Caribbean which traditionally depend on sugar exports.*

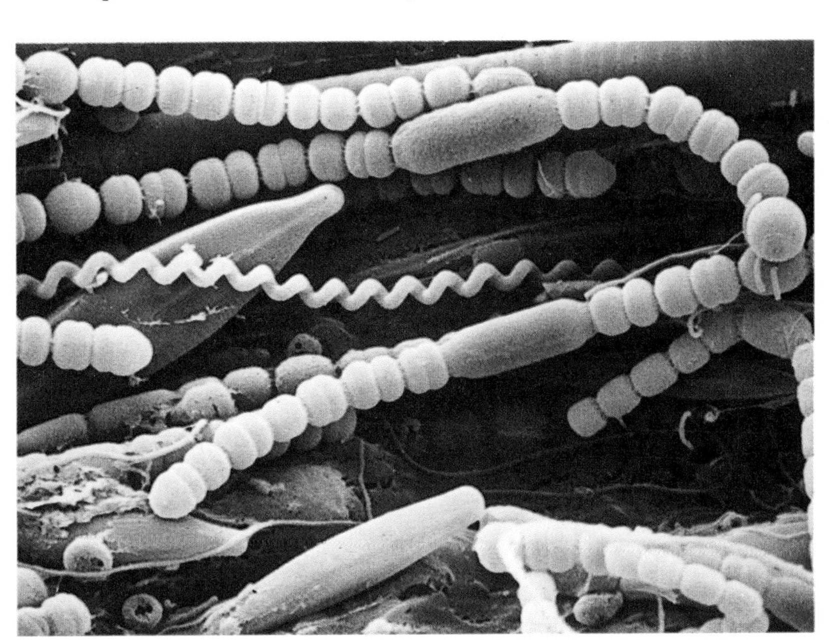

◀ *The blue-green alga, Anabaena, whose cells look like beads on a string, is a natural fertilizer factory. Using energy from the Sun, it converts nitrogen gas from the air into nitrogen that plants can use. This reaction takes place in specialized elongated cells called heterocysts. Indian researchers have used the organism to massively increase rice production in the paddy fields.*

Other technological breakthroughs are having a similar impact. Palm trees cloned in the United Kingdom from plant tissue and shipped to Malaysian plantations yield 30 percent more oil than the old stock. Many Malaysian rubber plantations are now being turned over to the more lucrative oil palms, and hundreds of thousands of workers in the labor-intensive rubber plantations are being laid off. Most of these are immigrants from Indonesia whose repatriation not only deprives Indonesia of an invisible export in the form of the workers' income but also adds to the large numbers of unemployed.

The increased volume of palm oil production has other repercussions too. Advances in both biotechnology and chemistry have made vegetable oils more "interconvertible" (◀ page 90). In other words, the properties of a particular oil (for example, boiling point, chemical reactivity, etc.) will no longer determine how that oil will be used. The food processing and chemical industries which both consume large volumes of vegetable oils can now choose which type of oil to use on the basis of cost alone. Soybean and palm oil have become the cheapest sources of vegetable oil, largely because the developed countries have used advanced techniques to improve the efficiency of oil production. The combination of interconvertibility and cheap production, both made possible by biotechnology, has changed the pattern of world trade in vegetable oils and will continue to do so.

A glut of palm oil lowers the price of all vegetable oil, threatening not only less-efficient palm oil producers, but also those traditional farmers growing coconuts and soybeans. There were 700,000 coconut farmers in the Philippines in 1985, most of them very poor and working areas of less than 5 hectares. As their markets failed, they found their income from coconuts fell by 40 percent of its value for the previous year. Tragically, those nations who are bearing the brunt of the biotechnology revolution are among the poorest and most vulnerable. Thailand, the Philippines, Sudan and many other developing countries rely heavily on exports of agricultural produce to the developed world to generate foreign exchange to fund their technological progress.

In the West farmers are not immune from the effects of biotechnology, either. European producers of sugar beet were unaffected by the development of HFCS because high tariffs imposed under the European Community's Common Agricultural Policy made it more expensive than sugar. But aspartame will undoubtedly erode the demand for beet, putting less-efficient farmers out of business.

The dairy industries of Europe and the USA will eventually have to contend with improved efficiency wrought by a better understanding of ruminant nutrition and genetic engineering of livestock. Milk production is already being improved with the growth hormone, bovine somatotropin (BST). BST is a natural protein of cows which is produced in pure form using genetically engineered microorganisms (◀ page 88). Since it can increase milk yields per cow by up to 30 percent, and consumption patterns will remain largely unchanged, smaller dairy herds or – more likely – fewer herds and fewer dairy farmers will be needed. Farmers whose animals have to struggle on poor land such as in mountainous areas would have the most to gain, but they could not afford the hormones. Once more, it will be the smaller, less-efficient farmers who lose out. The meat industry may be affected similarly by the use of growth hormones on cattle, pigs and sheep. Moreover, demand for meat might be reduced by biotechnological substitutes.

▶ Genes from elite organisms are big business in agriculture. The tanks contain the semen of "super cattle", preserved in liquid nitrogen for use in artificial insemination programs. Fertilized embryos are also stored in this way.

▶ Clones of superior oil palms destined for plantations in Malaysia. They are produced in laboratories in the United Kingdom using advanced techniques of plant-cell culture and regeneration. But they need Malaysian sunshine to develop into large trees which produce palm oil. Most plantations are owned by international firms and most oil is exported to industrial nations. Thus high technology can provide a more efficient way of transferring the Sun's energy from hot developing countries to temperate industrial ones.

Atlantic

▲ The Californian company Advanced Genetic Sciences was among the first to test genetically altered bacteria in the open environment. These strawberries were sprayed with "Frosban", an "ice minus" bacterium compound which protects them from frostdamage (♦ page 85). Earlier, AGS scientists contravened federal rulings and tested "Frosban" in an open roof garden. The company was fined.

► Vineyard owners have striven for centuries to produce grapes that give rise to wines of superior quality. That process is being given a more scientific basis by geneticists who have developed ways of growing vine plants in tissue culture. Each Petri dish may contain a different variety of plant which can be rapidly screened for resistance to infection or tolerance of certain soils.

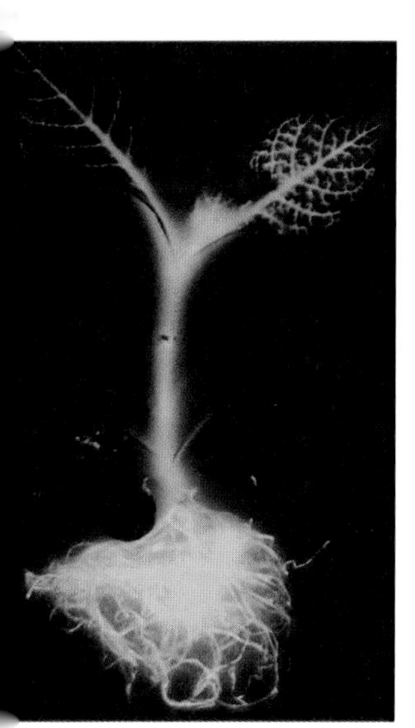

◀ *Generally, plants are grown only for one component: potato tubers, corn cobs or sunflower seeds, for instance. Genetic engineers need to ensure that a newly introduced gene will work in the right part of the plant. To show where genes are expressed, scientists have joined firefly genes responsible for light production to various genes in plants. As the gene produces its protein in the plant, a faint glow can be detected photographically. Thus genetic engineers can distinguish genes which produce proteins only in leaves (upper) from those needed in the most rapidly growing parts of the plant such as roots (lower).*

The new agricultural revolution

Biological products could be substituted for chemicals and other raw materials. Researchers at the Carlsberg company in Denmark have advanced the concept of the agricultural refinery, in the belief that agriculture and industry should interact more closely. Mechanical methods in agriculture only harvest cereal grain from a crop. The alternative is to garner the whole crop. Grain would be used not only for food and animal feed but would also supply the paper, food, textile fermentation and polymer industries with starch. Straw would be chopped into small pieces for making paper and particle board and the rest of the plant would be formed into pellets for fuel, animal feed and the chemical industry. Biotechnology would be responsible for many of the conversions required. Under this sort of scheme, crops would no longer be judged on a single property such as the baking quality of the flour from their grains. Instead, farmers could consider the relative yields of protein or starch or strawchips obtained from a variety of plants and process them according to the demand for the desired commodities. Genetic engineering and somoclonal variation (◀ page 80) could contribute crops with the necessary properties.

Control over the genetics of crops is also a key to commercial success in agriculture. Most of the development of new plants by plant-breeders and genetic engineers is carried out in the West, where much of the scientific knowledge of plants has accumulated. Traditionally, new varieties that had been developed were produced in bulk and distributed to farmers by small seed companies who specialized in catering for the needs of indigenous farmers. Over the past decade, however, this role has increasingly been taken over by multinational corporations who have bought up seed firms and invested heavily in biotechnology. These corporate giants are likely to develop high-value plants that are intended for farmers in western countries, rather than contribute to the production of varieties suited to agricultural practises in the developing world. A new variety of plant might, for instance, have artificial resistance to a herbicide (the herbicide would be sold to the farmer for weed control by another arm of the multinational company) and the ability to grow in the presence of a particular type of fertilizer (perhaps also supplied by the company).

In Japan and the United States, where companies can obtain patents for plant varieties, firms are sending expeditions across the world to gather different and unusual species of plants. Once these have been brought back, patented and grown in bulk, even the countries from which they had been taken would have to pay the price demanded. In response to this threat, groups of developing countries are now establishing their own stocks of plant germplasm in national and international plant-breeding centers. One of the major aims of these centers is to develop plants which are not reliant on intense inputs of fertilizer and pesticides.

In the West consumers will be partly cushioned against the new agricultural revolution brought about by the processing power of biotechnology. By gathering more information about the properties and functions of the produce we consume, scientists can identify the active ingredients and seek to obtain them elsewhere. Effectively, biotechnologists are dissecting produce and reconstituting it from the cheapest available sources. In the future, it will be difficult to tell if "meat" is of animal, vegetable or microbial origin. Milk might be hormone-plus, soya or regular. And the gas in automobiles could be refined from crude oil, vegetable oil or sugar cane.

With widespread embryo-testing in high-risk groups some genetic diseases might be eliminated in a generation

Information about our genes

Through DNA probe analysis, doctors will be able to tap into the intrinsic information contained in our genetic material. Some genetic diseases and the misery they cause could be virtually eliminated within a generation by widespread embryo testing in high-risk groups. Populations might become fitter and healthier as knowledge of our genes increases. Many people, however, feel uneasy about the prospect of widespread genetic screening. DNA sequence data like any other personal information, especially information that is retrievable from computer memory banks, can be misused.

Who needs to know about an individual's genetic profile? Should employers be told? For instance, deficiency in the gene that codes for the protein alpha-1-antitrypsin predisposes people to lung disease, especially in dusty or chemical-ridden environments. Trade unions have always opposed testing by employers on the grounds that it is discriminatory and absolves the company of the responsibility to clean up the workplace. What of the employee's obligations? Should someone who has been diagnosed positively for Huntington's disease, for instance, tell their employer? Not to do so might be deemed irresponsible since the earliest symptoms of the disease include irrationality and loss of muscle control. But a person who did tell might damage their career prospects.

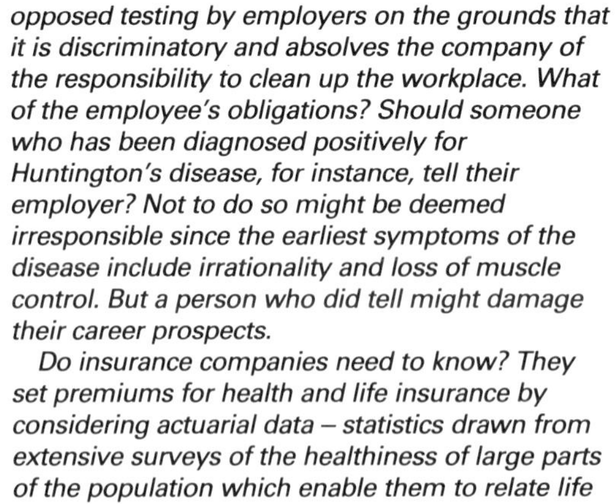

Do insurance companies need to know? They set premiums for health and life insurance by considering actuarial data – statistics drawn from extensive surveys of the healthiness of large parts of the population which enable them to relate life and health expectancy to age, gender, family history and habits (smoking, drinking, etc.) Genetic data could pinpoint more precisely the most likely causes of death or ill-health in an individual, and insurance companies could work out their premiums accordingly. People who would be likely to use the health services most often might be made to contribute more to their upkeep. On the other hand, people who inherit a genetic disability might be less likely to suffer sports injury or might die before old age becomes a burden on health resources.

For genetic analysis to be fair and effective, a huge number of factors would need to be assessed in screening. There is no current consensus as to whether a person's genetic inheritance should be allowed to influence other people's attitudes. On the one hand, equal-opportunities legislation bans discrimination on the basis of sex, race or disability. On the other, a family history of heart disease or stroke will influence life insurance premiums. Schools may want to know whether children have genetically determined learning disabilities. The military might wish to find out which candidates are predisposed to bravery, cowardice, or insubordination. The criminal justice and prison systems will certainly have an interest in knowing how genetic factors affect behavior. But would genetic disposition to crime be regarded as a mitigating factor or as further evidence against a hardened criminal?

Spreading the word

Biotechnology will enter everybody's lives. If you drink diet cola or beer it has already entered your digestive system. If you are a diabetic or a hemophiliac or have recently suffered a heart attack, it has probably entered your veins. If you are a farmer, it will be entering your fields as seed, inoculants and pesticides. If you live in a Third World country, it could take away your livelihood. Biotechnology generates information about life and transforms it into useful and often profitable products. The information itself, genetic sequence data for example, is sometimes that product. In the years to come, which of the products are used, and how, will depend on the judgment, not of scientists, but of administrators and lawmakers – and ultimately on the attitudes and opinions of ordinary people.

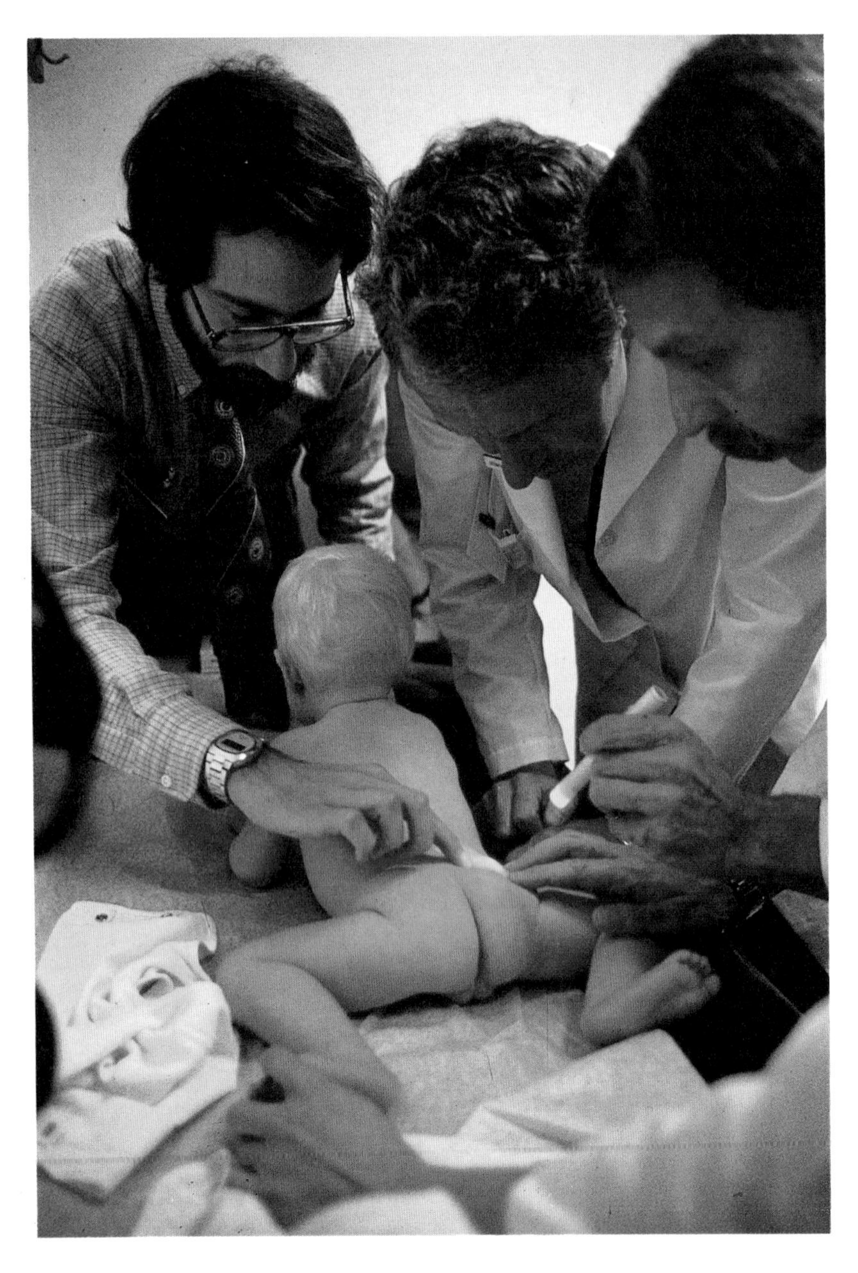

▼ *In modern healthcare the emphasis is on prevention rather than cure. A baby's DNA holds information on its future susceptibility to disease. DNA analysis should be a boon for medicine. But control over the use of genetic information will be needed to prevent its abuse.*

Glossary

Acquired Immune Deficiency Syndrome (AIDS)
Progressive damage to white blood cells caused by infection with human immune virus (HIV).

Amino acid
A basic chemical unit from which PROTEINS are synthesized by the body.

Antibiotic
A chemical produced by a MICROORGANISM and used as a drug to kill or inhibit the growth of other microorganisms.

Antibody
A defensive substance produced by the immune system to neutralize or help destroy a specific foreign substance.

Antigen
A foreign substance that provokes the body to produce ANTIBODIES.

Auto-immune disease
A disease in which the body rejects some of its normal tissues and mobilizes the immune system against them.

Bacteria (singular bacterium)
A large and varied group of MICROORGANISMS, classified by their shape and staining ability. They live in many environments; only a few are harmful.

Bacteriophage
A VIRUS that attacks BACTERIA.

Biological control
The use of one species (usually a predator or parasite) to control the population of another.

Biotechnology
The use of MICROORGANISMS usually for industrial purposes.

Catalyst
A substance which increases the speed of a chemical reaction but is itself chemically unchanged at the end of the reaction.

Chromosomes
Strictly speaking, the structures that carry the genetic information in eukaryotic cells only. They are made up of DNA and proteins and have a complex structure. (The much simpler genetic material of prokaryotic cells, consisting of a loop of DNA or RNA with just a few associated proteins, is sometimes referred to as a bacterial chromosome.)

Clone
A group of genetically identical organisms or cells, having all derived from a single ancestor. See also GENE CLONING.

Cytoplasm
The cell's contents but excluding in eukaryotic cells, the region inside the nucleus.

Differentiate
Of cells and tissues, to take on different forms during development or regeneration.

Diversity
A measurement of the richness of species in a given area.

DNA
Deoxyribonucleic acid; its structure contains the blueprint that contains genetic information.

Ecology
The study of the relationships and the interactions of living organisms, with each other and with the physical world.

Embryo
In humans, the developing baby in the womb from conception to the end of the tenth week.

Energy
The ability to do work. It may take many forms, such as light, heat and chemical.

Enzyme
A PROTEIN which is a CATALYST of biochemical reactions. There are many different kinds, each kind directly promoting only one or a very limited range of reactions.

Fetus
The developing baby, from the tenth week after conception to birth.

Gene
A unit of hereditary information. Each gene produces a single polypeptide chain.

Gene cloning
The reproduction of a gene using genetic engineering techniques. See also RESTRICTION ENZYME and VECTOR.

Gene machine
Popular name for oligonucleotide synthesizer, a computer-controlled instrument that can manufacture any DNA sequence, by chemically adding bases to a growing chain of DNA.

Gene therapy
The treatment of genetic disease by treating a patient's genes. Somatic gene cell therapy is performed on the blood cells and any changes in the genes will not be passed on. Germline cell gene therapy, performed on the reproductive cells, could, in principle, both cure a genetic defect and prevent it from being inherited.

Genetic engineering
Methods of altering an organism's characteristics by direct manipulation of the DNA, usually by inserting foreign genes; technical name, recombinant DNA technology.

Genetics
The study of heredity.

Genome
The total genetic complement of an organism.

Germline gene therapy
The proposed addition of extra genes to fertilized eggs to cure genetic defects for all future generations.

Heredity
The passing of genetic characteristics from parents to children.

Hormone
A chemical secreted by an endocrine gland which has a specific effect on a target cell in another part of the body.

Hybrid
Offspring of parents which are not genetically identical.

Immunity
A state of resistance to an infection, through the existence of ANTIBODIES and immune cells specifically able to attack the MICROORGANISMS responsible.

In vitro
In glass; outside the living body and within an artificial environment such as a test tube.

Metabolic rate
The general level of METABOLISM in an animal, in terms of the amount of energy expended (usually excluding that involved in movement etc). It is calculated by measuring the amount of oxygen consumed.

Metabolism
The chemical process occurring within an organism, including the production of PROTEINS from AMINO ACIDS, the exchange of gases in respiration, the liberation of energy from goods and innumerable other chemical reactions.

Microorganism
Organisms of microscopic or ultramicroscopic size, such as bacteria, some fungi, viruses.

Molecule
An entity composed of atoms linked by chemical bonds and acting as a unit; its composition is represented by its molecular formula.

Mutation
A structural change in a GENE which may give rise to a new heritable characteristic if it occurs in one of the germ cells.

Neuron
A nerve cell.

Node
In plants, the point at which a leaf or leaves develop from the stem.

Oncogene
A GENE carried by a VIRUS that is involved in transforming normal cells to cancerous ones orginally a section of DNA picked up by a virus.

Organic molecules.
Complex carbon-containing molecules.

Organism
Any living thing.

Parasite
An organism that lives in or on another organism (the host) and is metabolically dependent on it.

Pathogen
An organism that produces disease.

Peptide
A chain formed from two or more AMINO ACIDS. When several amino acids are involved, the chain is known as a POLYPETIDE. These names are applied to a partially synthesized PROTEIN, to the fragments obtained during the digestion of proteins, to short chains synthesized in the laboratory, or to single chains that form part of a larger protein molecule.

pH
A numerical measurement, from 0-14, of the acidity or alkalinity of liquids. A pH of 7.0 is neutral, less than 7.0 is acidic, more than 7.0 is alkaline.

Phage
See BACTERIOPHAGE.

Photosynthesis
The synthesis of organic compounds, primarily sugars, from carbon dioxide and water using sunlight as the source of energy, and chlorophyll, or some other related pigment, for trapping the light energy.

Polypetide
See PEPTIDE

Protein
A complex bio-molecule, made up of one or more chains of AMINO ACIDS. Where made of several chains, each of these is known as a POLYPETIDE chain.

Protoplasm
The contents of the cell. In eukaryotes it is divided into the nucleus and CYTOPLASM.

Respiration
Strictly speaking, the breakdown of food molecules in the presence of oxygen, a biochemical process which releases energy.

Restriction enzyme
An ENZYME that always cuts a DNA molecule in the way, by recognizing a sequence of bases; used in GENE CLONING.

RNA
Ribonucleic acid; a single-stranded nucleic acid that cooperates with DNA for PROTEIN synthesis.

Somoclonal variation
A method of cloning plants by growing plants from single cells, breaking the plants down, in turn, into single cells and growing these to produce identical plants.

Spore
A rather general term applied to a great variety of small reproductive or resting bodies. In plants, algae, fungi and slime molds, the spore is a simple reproductive body, usually consisting of a single cell with a protective coat. The term spore is also applied to structures that have no reproductive role, such as bacterial spores.

Symbiosis
A close relationship between organisms of different species.

Synapse
The junction between the processes of adjacent NEURONS, across which nerve impulses are carried by transmitter substances.

Tumor
A growth of excess tissue due to abnormal cell division.

Vaccination, vaccine
Orginally, the introduction of matter from cowpox pustules to lessen the danger of catching smallpox; by extension, vaccines are attenuated forms of disease organisms used to confer immunity.

Vector
A carrier molecule in which a new fragment of DNA is inserted into the host cell for the purpose of GENE CLONING. Vectors include plasmids, cosmids and bacteriophages.

Virus
The smallest form of living organism, dependent on living cells for replication.

Vitamin
A substance which is essential for life and which the body cannot synthesize, so it must be present in diet.

Units of Measurement

In general the International System of Units – SI units – is used throughout this book. This system is founded upon seven empirically defined *base units* (eg ampere), which can be combined, sometimes with the assistance of two geometric *supplementary units* (eg radian) to yield the *derived units* (eg cubic meter) which together with the base units constitute a coherent set of units capable of application to all measurable physical phenomena. Some of the derived units have special names (eg volt). For the sake of convenience smaller and larger units, the multiples and submultiples of the SI units, can be formed by adding certain prefixes (eg milli-) to the names of the SI units. In any instance only one prefix can be added to the name of a unit (thus nanometer, not millimicrometer). Of the other units in common scientific use, it is recognized that several (eg hour) will continue to be used alongside the SI units, although combinations of these units with SI units (as in kilowatt hour) are discouraged. However, other units (eg angstrom unit) are redundant if the International System is fully utilized, and it is intended that these should drop out of use.

Base units

Quantity	Unit	Symbol
length	meter	m
mass	kilogram	kg
time	second	s
electric current	ampere	A
temperature	kelvin	K
luminous intensity	candela	cd
amount of substance	mole	mol

Derived units

Quantity	Name	Symbol
frequency	hertz	Hz
force	newton	N
work, energy	joule	J
power	watt	W
pressure	pascal	Pa
quantity of electricity	coulomb	C
potential difference	volt	V
electric resistance	ohm	Ω
capacitance	farad	F
conductance	siemens	S
magnetic flux	weber	Wb
flux density	tesla	T
inductance	henry	H
luminous flux	lumen	lm
illuminance	lux	lx

Supplementary units

Quantity	Name	Symbol
plane angle	radian	rad
solid angle	steradian	sr

Non-SI units in common use

Quantity	Name	Symbol
plane angle	degree	°
plane angle	minute	′
plane angle	second	″
time	minute	min
time	hour	h
time	day	d
volume	liter	l
mass	tonne	t
energy	electronvolt	eV
mass	atomic mass unit	u
length	astronomical unit	AU
length	parsec	pc

Index notation

Very large and very small numbers are often written using powers of ten. The American system in the table below names large numbers according to the number of groups of three zeros which follow 1,000 when they are expressed in numerals – eg 1 billion (bi- meaning two) is 1,000 followed by two groups of three zeros.

Name	Numeral	Value in powers of ten	SI prefix	SI symbol
one	1	10^0	—	—
ten	10	10^1	deka	da
hundred	100	10^2	hecto	h
thousand	1,000	10^3	kilo	k
million	1,000,000	10^6	mega	M
billion	1,000,000,000	10^9	giga	G
trillion	1,000,000,000,000	10^{12}	tera	T
tenth	0.1	10^{-1}	deci	d
hundredth	0.01	10^{-2}	centi	c
thousandth	0.001	10^{-3}	milli	m
millionth	0.000001	10^{-6}	micro	μ
billionth	0.000000001	10^{-9}	nano	n
trillionth	0.000000000001	10^{-12}	pico	p

Abbreviations not already given above

A	mass number
Å	angstrom unit
AC	alternating current
AF	audio frequency
asb	apostilb
atm	atmosphere
AW	atomic weight
bhp	brake horse power
bp	boiling point
Btu	British thermal unit
C°	centigrade degree
°C	degree Celsius
Cal	see Kcal
cal	calorie
ccp	cubic close-packed
CGS	centimeter-gram-second (system)
Ci	curie
dB	decibel
DC	direct current
EHF	extremely high frequency
emf	electromotive force
emu	electromagnetic unit
esu	electrostatic unit
f	femto-
fcc	face-centered cubic
FM	frequency modulation
G	universal constant of gravitation
g	gram
g	acceleration due to gavity
Gb	gilbert
gr	grain
Gs	gauss
h	Planck constant
ha	hectare
hcp	hexagonal close-packed
HF	high frequency
hp	horse power
IF	intermediate frequency
ir	infrared
k	kilo-
k	Boltzmann constant
kcal	kilocalorie
kgf	kilogram force
L	lambert
LW	long wave
ly	light year
M	mega-
mbar	millibar
MF	medium frequency
MKSA	meter-kilogram-second- ampere (system)
mp	melting point
mya	million years ago
MW	medium wave
MW	molecular weight
Mx	maxwell
$\mathcal{N}$	Avogadro number, neutron number
nmr	nuclear magnetic resonance
Oe	oersted
O.N.	oxidation number
P	poise
pH	hydrogen ion concentration
ph	phot
ppm	parts per million
R	röntgen
R	universal gas constant
°R	degrees Rankine or (Reaumur)
rad	radian
rd	rad (dose)
rpm	revolutions per minute
sb	stilb
sg	specific gravity
SHF	superhigh frequency
sr	steradian
STP	standard temperature and pressure
subl	sublimation point
UHF	ultrahigh frequency
uv	ultraviolet
VHF	very high frequency
yr	year
$\mathcal{Z}$	atomic number

Index

Page numbers in *italics* refer to illustrations, captions or side text.

A

B

C

D

E

F

G

Credits

Key to abbreviations: CD Chemical Designs Ltd; FSP Frank Spooner Pictures; MEPL Mary Evans Picture Library; NHPA Natural History Photographic Agency; OSF Oxford Scientific Films; SPL Science Photo Library; TCL Telegraph Colour Library; b bottom; bl bottom left; br bottom right; c center; cl center left; cr center right; t top; tl top left; tr top right; l left, r right.

5 SPL/Dr D. Fawcett **6–7** SPL/J. Burgess **6r** SPL/CNRI **8l** MEPL **8r** CD **8–9** Magnum/E. Hartmann **10, 11, 12, 13** CD **14** SPL/Science Source **15, 16, 17, 18, 19, 20, 21** CD **24l** Ted Spiegel **24r** SPL/Dr T. Brain **25, 26b** Ted Spiegel **26–27** Magnum/E. Hartmann **28l** SPL/Dr G. Murti **28r** SPL/Dr L. Simon **29t** Amersham International plc **29b** Magnum/E. Hartmann **30** Ted Spiegel **30–31** SPL/Petit Format/D. Schouvaert **31t** SPL/P.A. McTurk, University of Leicester and D. Parker **31r** FSP **32l** OSF/R. Blythe **32r** NHPA/I. Polunin **33** Cetus Corporation **34** Celltech Ltd **35l** Ted Spiegel **35r** Cetus Corporation **36t** FSP **36b** Mail on Sunday **37l** Ted Spiegel **37r** OSF/R. Packwood **38–39** OSF/P.J. DeVries **39** Australasian Nature Transparencies/Natfoto **40l** CD **40–41** FSP **42–43t** SPL/CNRI/Professor Luc Montagnier, Institute Pasteur **43b** Ann Ronan Picture Library **43br** TASS **44** SPL/Synaptek **45l** SPL/R. Eagle **45r** Novo **46** NHPA/G. Bernard **47t** OSF/M. Fogden **47b, 49** CD **50** NASA **51t** CD **51b** SPL/Professor Max Perutz, MRC Laboratory of Molecular Biology **52** CD **53l** Rainbow/D. McCoy **53r, 54, 55, 56, 57, 58, 59, 60** CD **61** Gist-Brocades **62–63** Image Bank/M. Melford **63t, 63b** Gist-Brocades **64–65** OSF/G.A. Maclean **65t** Rowntree **65r** Unilever/G. Talbot/Dr Padley **66–67** SPL/CNRI **66l** SPL/Division of Computer Research Buntrock **82b** Gist-Brocades **83** Celltech Ltd., **84–85t** SPL/Gene Cox **84–85b** SPL/Dr J. Burgess **85** Ted Spiegel **86** SPL/CNRI/Tektoff-RM **86** inset FSP **87** Ted Spiegel **88l** Agriculture and Fisheries Research and Technology, National Institute of Health **68** Jerry Mason **68–69** Celltech Ltd., **69r** SPL/Dr J. Burgess **70** Gist-Brocades **71t** Robert Harding/J.F.E. Russell **71b** Bruce Coleman Ltd./W. Lankinen **72l, 72r, 73b** Ted Spiegel **73t** Celltech Ltd, **74–75** NHPA/D. Woodfall **75tr** SPL/Dr. J. Burgess **75br** Cornell University **76** CD **77** Cetus Corporation **78l** Ted Spiegel **78r** Munton and Fison **79** Twyfords **80l** LH Fermentation **81tl** Ted Spiegel **81tr** NHPA/R. Knightsbridge **81–82b** NHPA/I. Polunin **82t** Anthony Blake Library/E. Council **88r, 89** Ted Spiegel **90** Agriculture and Fisheries Research Council **91t, 91b** SPL/CNRI **92l** Ted Spiegel **92r** CD **93t** SPL/CNRI/Tektoff **92–93b** SPL/CNRI **94** Hutchison Library **94–95** FSP **95r** SPL/H. Morgan **96** Gist-Brocades **97t** Hutchison Library/Dr N. Smith **97b** CD **98b** SPL/H. Morgan **98t, 99l, 99tr, 99br, 100tr** Ted Spiegel **100tl, 101** Damon Biotech **100–101b** Cetus **102–103t** Illustrated London News **102br** Melody Maker **103** SPL/Bill Longcore **104l** SPL/M. Dohrn/IVF Unit Cromwell Hospital **104r** SPL/CNRI **105l** SPL/R. Hutchings **105r** SPL **106l** Cetus Corporation **106r, 107l** Ted Spiegel **107b** CD **108l** SPL/Astrid/Kage **108tr, 109t** Ted Spiegel **108–109** Cetus Corporation/J.C. Hunter Cerera **110t, 110b** National Medical Slide Bank **111l, 111r, 112l** Ted Spiegel **112–113** Magnum/F. Scianna **113b** Cetus Corporation **114l** FSP **114r** Hutchison Library/V. Thapat **115t** News and Publications Service, Stamford University **115b, 116r** Ted Spiegel **116l** Genentech Inc. **117** FSP **118t** CD **118–119b** Lawrence Clarke **119** SPL/Dr T. Brain **120–121** Unilever **122t, 122b, 123t, 123b** FSP **124** Magnum/E. Hartmann.

Index Barbara James and John Baines
Production Joanna Turner, Clive Sparling
Typesetting Anita Rokins
Media conversion Peter MacDonald